Analyzing Organic and Fairtrade Certification Schemes: Participation and Welfare Effects on Small-Scale Farmers in Coffee Value Chains

Analyzing Organic and Fairtrade Certification Schemes: Participation and Welfare Effects on Small-Scale Farmers in Coffee Value Chains

Tina Beuchelt

Cuvillier Verlag Göttingen 2012

Bibliografische Information der Deutschen Nationalbibliothek

Die Deutsche Nationalbibliothek verzeichnet diese Publikation in der Deutschen Nationalbibliographie; detaillierte bibliographische Daten sind im Internet über http://dnb.d-nb.de abrufbar.

1. Aufl. - Göttingen: Cuvillier, 2012

  Zugl.: Hohenheim, Univ., Diss., 2012

  ISBN 978-3-95404-119-0

D100
Dissertation University of Hohenheim, Institute of Agricultural Economics and Social Sciences in the Tropics and Subtropics, Chair: Rural Development Theory and Policy, 2012.

This thesis was accepted as a doctoral dissertation in fulfillment of the requirements for the degree "Doktor der Agrarwissenschaften" by the Faculty Agricultural Sciences at University of Hohenheim on 25 May 2011.

Date of oral examination: 14 July 2011

Examination Committee:
Supervisor and Review:	Prof. Dr. M. Zeller
Co-Reviewer:	Prof. Dr. A. Bellows
Additional Examiner:	Prof. Dr. S. Dabbert
Vice-Dean and Head of Committee:	Prof. Dr. A. Fangmeier

© CUVILLIER VERLAG, Göttingen 2012

  Nonnenstieg 8, 37075 Göttingen

  Telefon: 0551-54724-0

  Telefax: 0551-54724-21

  www.cuvillier.de

1. Auflage 2012

Gedruckt auf säurefreiem Papier.

  ISBN 978-3-95404-119-0

Smallholder
Coffee
Production in
Nicaragua

"Coffee pays for everything, from the shoes to the top [...].
Everything comes from the same coffee. That is why it never
gives us enough to improve our living."
Conventional coffee smallholder

Contents

List of Figures

List of Tables

Abbreviations

CONV	Conventional Cooperative
FOB	Free on Board
ICE	Intercontinental Exchange
ICO	International Coffee Organization
IFOAM	International Federation of Organic Agriculture Movements
FLO	Fair Trade Labelling Organization
NGO	Non-Governmental Organization
ORG	Organic Certified Cooperative
OFT	Organic-Fairtrade Certified Cooperative
SWOT	Strengths, Weaknesses, Opportunities, and Threats
WFTO	World Fair Trade Organization

The exchange rate for Nicaraguan Córdoba against US\$ was 1US\$ = 18.96C\$ as of 31 August 2007.

Executive Summary

Organic and Fairtrade certified coffees have become very popular among socially, environmentally and health conscious consumers in recent years. As consumers pay higher prices for these certified coffees, it is commonly assumed that, compared to conventional coffee, better producer prices are paid and that higher shares of the added value in consuming countries trickle down to the producers. Coffee certifications are thus supposed to benefit the coffee producers. Coffee is an important export good for many developing countries. The majority of global coffee production comes from around 20-25 million smallholder families in developing countries. As individual certifications are too expensive smallholders have to participate in farmer organizations, e.g. cooperatives, in order to access cheaper group certification. Governments and international donors support coffee certification schemes and assume that these link farmers to high-value markets, increase producers' incomes, change power and information asymmetries in value chains, and contribute to poverty reduction. Yet, there is only weak empirical evidence that justifies this support. There are few quantitative studies which applied random sampling techniques, and analyzed the effects of certification schemes in regard of gross margins, profits, income shares and poverty levels of certified smallholder coffee producers. The role of cooperatives for the success of certification schemes has been neglected by research. The available studies have methodological limitations, for example they are based on qualitative methods only, include no more than one cooperative or one certification standard, or cooperatives are non-randomly sampled.

This research seeks to fill the identified knowledge and methodological gaps. Through a combination of qualitative and quantitative research, the production and marketing strategies of small-scale coffee producers in northern Nicaragua are compared based on producers that are organized in conventional, organic, and Organic-Fairtrade certified cooperatives. The analysis addresses (i) the smallholders' household level and (ii) the organizational and institutional level with regard of the cooperatives and respective coffee value chains. The study aims at, first, identifying the socio-economic costs and benefits of participation in organic and Organic-Fairtrade certified coffee chains with respect to level of coffee and household incomes as well as household poverty. Second, it is examined which role the farmer organizations, their respective business models and upgrading strategies, play for the success or failure of certification schemes. Third, the integration of coffee farmers and their cooperatives into the coffee value chain, the structure and functioning of the value chains and the value adding effect of certification is examined.

The survey was conducted in the northern Nicaragua departments Madriz, Nueva Segovia, and Matagalpa on coffee farms situated between 900m and 1300m a.s.l. The coffee of all farmers was classified as 'Strictly High Grown'; the species is *Coffea Arabica*. The sample design ensured that the research region was homogeneous with respect to living conditions, socio-economic level, as well as coffee growing characteristics driving performance of coffee farmers. After having randomly selected the cooperatives, 327 coffee producing households were also randomly selected and surveyed with a structured questionnaire. Qualitative data collection consisted in total of 58 key-person interviews, 67 semi-structured farmer interviews and 24 focus group discussions with coffee farmers. The primary data was collected during two research stays in 2007 and 2008.

This research analyzes gross margins, accounting and economic profits of coffee production. The household income is measured and a poverty headcount index elaborated. Principal component analysis is used to determine current relative poverty levels and the development of relative poverty over time. A SWOT analysis identifies the strengths, weaknesses, opportunities, and threats of cooperatives. Through a value chain analysis information on the actors, power and information flows as well as price shares is gained. For identifying the farmers' experiences with coffee certification schemes, a thematic analysis is applied to the qualitative data by developing an individual code system for data-reduction.

In the research region, the coffee yields of conventional and certified coffee smallholders are usually 40% to 50% lower than national average due to limited maintenance activities and inadequately managed coffee plantations. Highest yields (on average around 480kg/ha) are achieved by organic producers but yield levels vary, like for conventional and Organic-Fairtrade certified producers, between the cooperatives (ranging from 293kg/ha to 516kg/ha). In comparison to conventional prices, Organic-Fairtrade certified coffee achieved on average 11% and organic coffee 8% higher farm-gate prices; price differences between cooperatives also exist. Organic production processes require fewer purchased inputs but are more laborious. Due to constrained availability of family labor, additional labor has to be hired which offsets saved input costs. The higher prices of certified coffees compensate for production costs but fail to increase per hectare gross margins and profits in the case of Organic-Fairtrade farmers compared to conventional produces. Due to higher yield levels, organic producers experience an increase in per hectare gross margins and profits. They have with 328US$/ha a significantly higher economic profit than Organic-Fairtrade farmers (147US$/ha) and conventional farmers (191US$/ha). Yet, as they tend to have smaller coffee areas and larger family sizes, the increase in gross margins does not result in improved per capita net coffee incomes for organic certified producers compared to the other groups. Also Organic-Fairtrade certified producers do not have higher per capita net coffee incomes than conventional producers.

Among organic and Organic-Fairtrade certified producers, a higher share of households is grouped below the extreme poverty line than among conventional producers (45% compared to 30%) – which means that they cannot cover their food requirements. Between 60% and 70% of conventional and certified coffee producers are below the national poverty line. Using principal component analysis to investigate several dimensions of poverty and their development over time, it was found that over a period of ten years, organic certified producers became relatively poorer. In the year 1997, all groups had similar relative poverty levels. The Organic-Fairtrade certified producers first improved their relative poverty status during the coffee crisis (in 2002) and were relatively better off than conventional producers. Since then, the relative poverty levels of Organic-Fairtrade producers deteriorated compared to conventional producers.

Irrespective of whether farmers were certified or not, Nicaragua's coffee smallholders face two to three months of food shortages per year during which they seek off-farm employment, and apply for formal and informal credits. In many cases the credit is used for immediate consumption needs, like food or medicine, and only partially invested in the farm. Consequently, harvested yields stay low, leading to low incomes and new credit requirements. When farmers are financially illiterate or requested higher credits than their payment capacity, they are likely to enter a vicious cycle of indebtedness.

Each cooperative has a unique business model; they differ, for example, in member size, functions and services, internal organization, and financial characteristics. Despite their different business models the cooperatives often choose the same upgrading strategies as other cooperatives mainly certification, quality, and own processing. The analysis of strengths, weaknesses, opportunities, and threats (SWOTs) showed that the cooperatives have certain SWOTs in common but there are also cooperative specific SWOTs. The common strength of the cooperatives is the quality potential of the region. The common weaknesses relate to the lack of credit access, a weak extension system, and weak rural infrastructure. The common threats of the cooperatives are high competition among national coffee buyers and cooperatives, corruption and mismanagement, and, according to the qualitative interviews, increasing microclimatic variations and unreliable rainfall patterns. The common opportunities range from more horizontal coordination to reduce transaction costs to share certificates acknowledging the members' possessions in the cooperative and increased transparency about deductions on payments. Qualitative evaluation indicated no obvious association between the coffee certification strategy of farmers/their cooperative and the coffee gross margins farmers obtained. The upgrading strategies of cooperatives, the strengths and weaknesses as well as the amount of coffee-related services, which the cooperative offers to producers, tend to be more related to coffee gross margins than the organic or Organic-Fairtrade certification.

Farmers are found to have no bargaining power over prices irrespective of the value chain, while certified cooperatives have limited bargaining power towards their buyers compared to cooperatives in the conventional chain. Power is unequally distributed between buyers and sellers of coffee in all chains. The quantity and quality of information flows depends on the cooperative and value chain model. Information asymmetries are fewer in certified chains; yet this also depends on the cooperative. Organic-Fairtrade certified value chains tend to have more and smaller-sized actors, especially in consuming countries, compared to the conventional chain. This increases transaction costs in the certified value chains and thus leads to substantially lower producers' share of the final coffee retail price (8%-15% in certified chains compared to 24%-34% in conventional chains).

The presented results depend strongly on each cooperative and there are large variations within the organic and Organic-Fairtrade certified cooperatives. It can be concluded that higher farm-gate coffee prices do not lead necessarily to higher per capita net coffee and household income, as yield levels, production costs, family and land size, as well as labor availability play important roles.

Organic or Organic-Fairtrade certification as an upgrading strategy seems only then successful when the business model of a cooperative, its strengths, weaknesses, and other upgrading strategies are supportive. Given the
constraints mentioned above, a well functioning cooperative is a necessary but not sufficient condition. This was shown by the example of one well run Organic-Fairtrade certified cooperative with low gross margins showed.

The main causes of continuing poverty among smallholder coffee growers in northern Nicaragua seem not the lack of market access or so-called 'unfair' trading conditions. Based on the qualitative analysis, reasons for poverty are lack of entrepreneurial and management skills of farmers and cooperative staff, financial illiteracy and indebtedness of farmers as well as a very weak rural infrastructure. Based on the quantitative results potential reasons for poverty are low yield and productivity levels, land and labor constraints. Certification schemes do not address or are able to solve these problems. Prices for certified coffee cannot compensate for low productivity, land or labor constraints.

Therefore, certification schemes can only be part of a viable development policy for poor small-scale farmers in northern Nicaragua; the production, infrastructural, organizational and institutional problems mentioned above require even more attention from policy makers. It is recommended that policies, which aim at increasing smallholder coffee incomes through upgrading, should focus apart from production aspects on the institutional context of smallholders and their cooperatives. Regarding coffee production, policies should address coffee yield levels, for example through research investments in improved, stress-tolerant and locally adapted varieties to encounter the microclimatic variations. Coffee quality in the region should be further strengthened by a supportive coffee sector strategy at the national level, which should include a national coffee institute or federation like in Colombia or Costa Rica. This should be accompanied by investments in rural infrastructure. It is recommended to establish an efficient extension system which also addresses the entrepreneurial skills of farmers. This could be also in form of facilitating the establishment of extension associations which could operate regionally and be financed by their members' contribution.

In order to better link farmers to (high-value) markets and to increase their income, it is recommended to focus more on the structure and functioning of producer organizations and their respective value chains. Business and strategic advice to cooperatives is necessary, as cooperative leaders and staff are not fit for international markets, in which they have to act. A banking system which also provides credits to cooperatives (at market interest rates and lending conditions) would reduce the reliance and dependence on exporters or international credit providers and could ease liquidity constraints of cooperatives. An obligatory annual external auditing of cooperatives, like it exists in other countries, is considered to be important to reduce mismanagement of a cooperative. It will also increase the creditworthiness of cooperatives for banks.

Trade, processing, and marketing efficiencies in the organic but especially in the Fairtrade value chains in consuming countries need to be improved in the alternative trade sector with its many small profit or non-profit enterprizes and organizations. These actors could consolidate to exert economies of scale and reduce their transaction costs. Consolidation is certainly a new way of thinking in the alternative trade sector but could effectively contribute to improve farmers' shares of retail prices and raise farm-gate coffee prices.

Zusammenfassung

Biologisch (ökologisch) zertifizierte und fair gehandelte (Fairtrade) Kaffees sind in den letzten Jahren bei umweltbewussten, gesundheitsbewussten und sozial engagierten Konsumenten sehr populär geworden. Da die Konsumenten für zertifizierte Kaffees höhere Preise zahlen, wird allgemein angenommen, dass, im Vergleich zu konventionellem Kaffee, die Kaffeebauern höhere Preise erhalten und ein größerer Anteil an der Wertschöpfung die Kaffeebauern erreicht. Darauf basiert die allgemeine Annahme, dass Zertifizierung von Kaffee sich auf die Lebensumstände der Bauern positiv auswirkt.

Kaffee ist ein wichtiges Exportgut für viele Entwicklungsländer. Die Mehrheit der weltweiten Kaffeeproduktion stammt von ca. 20-25 Millionen Kleinbauernfamilien in Entwicklungsländern. Da eine individuelle Zertifizierung für Kleinbauern unerschwinglich ist, müssen sie Mitglieder in Bauernorganisationen, z.B. Genossenschaften, werden, um Zugang zu der günstigeren Gruppenzertifizierung zu erhalten. Regierungen und internationale Geldgeber unterstützen die Kaffee-Zertifizierung in der Hoffnung, dass diese Bauern Zugang zu Märkten mit hoher Wertschöpfung biete, das Einkommen der Bauern erhöhte und so zur Armutsreduzierung beitragen könnte sowie Macht- und Informationsasymmetrien in Wertschöpfungsketten verringern würde. Die empirische Basis, die diese Unterstützung rechtfertigt, ist schwach. Es gibt nur sehr wenige quantitative Studien, die auf zufällig ausgewählten Stichproben beruhen, und die die Auswirkungen von Zertifizierungen auf Deckungsbeiträge, Profite, Einkommensanteile und das Armutsniveau von kleinbäuerlichen Kaffeebauern analysieren. Die Rolle der Genossenschaften für den Erfolg von Zertifizierungen wurde bisher von der Forschung vernachlässigt. Zusätzlich haben die verfügbaren Studien oft methodische Schwächen, zum Beispiel basieren sie nur auf qualitativen Methoden, betrachten nur eine Genossenschaft oder einen Zertifizierungsstandard, oder die Genossenschaften/Bauernorganisationen sind nicht zufällig ausgewählt.

Diese Studie strebt an, diese methodischen und wissenschaftlichen Lücken zu decken. Mit einer Kombination von quantitativer und qualitativer Forschung werden die Produktions- und Vermarktungsstrategien von kleinbäuerlichen Kaffeebauern im Norden Nicaraguas verglichen, welche in konventionelle, biologische sowie biologisch und Fairtrade (Bio-Fairtrade) zertifizierte Genossenschaften organisiert sind. Es werden zwei analytische Ebenen behandelt: (i) Die kleinbäuerliche Haushaltsebene und (ii) die organisatorische und institutionelle Ebene mit Blick auf die Genossenschaften und ihre jeweiligen Wertschöpfungsketten. Die vorliegende Studie strebt an, zuerst die sozio-ökonomischen Kosten und Nutzen der Teilnahme an biologisch und Bio-Fairtrade zertifizierten Kaffeewertschöpfungsketten hinsichtlich des Kaffee- und Haushaltseinkommen sowie des Armutsniveaus zu identifizieren. Als zweites Ziel wird analysiert, welche Rolle die Genossenschaften, ihre Geschäftsmodelle und Kaffee-Veredelungsstrategien ('Upgrade-Strategien') für den Erfolg von Kaffeezertifizierungen spielen. Als drittes wird die Integration der Kaffeebauern und ihrer Genossenschaften in die Kaffeewertschöpfungskette, die Struktur der Wertschöpfungskette und der Anteil der kleinbäuerlichen Wertschöpfung untersucht.

Die Studie basiert auf einer Primärdatenerhebung nicaraguanischer Kaffeekleinbetriebe, die auf einer Höhe von 900m bis 1300m über dem Meeresspiegel in den nördlichen Departements Madriz, Nueva Segovia, und Matagalpa lagen. Der Kaffee aller Bauern fiel in die Qualitätskategorie 'Strictly High Grown', die Kaffeeart war *Coffea Arabica*. Für den Stichprobenplan war sichergestellt, dass die Forschungsregion homogen in Bezug auf Lebensbedingungen und das sozio-ökonomische Lebensniveau sowie homogen bezüglich der Anbaucharacteristiken von Kaffee war, welche die Effizienz der Bauern bestimmen. Nach Auswahl der Kaffeevermarktungsgenossenschaften durch eine Zufallsstichprobe wurden von diesen ebenfalls per Zufallsstichprobe 327 Haushalte von Kaffeebauern ausgewählt und mit einem standardisierten Fragebogen interviewt. Die qualitative Datensammlung bestand aus insgesamt 58 Interviews mit Schlüsselpersonen, 67 Leitfaden-basierten Interviews und 24 Fokusgruppen-Diskussionen mit Kaffeekleinbauern. Die Datensammlung fand während zwei Forschungsaufenthalten in 2007 und 2008 in Nicaragua statt.

Diese Forschungsarbeit analysiert Deckungsbeiträge, Gewinne und kalkulatorische Betriebszweigergebnisse für die Kaffeeproduktion. Das Haushaltseinkommen wird erfasst und ein Armutsindex erstellt. Mittels einer Hauptkomponentenanalyse wird das gegenwärtige relative Armutsniveau bestimmt und die Entwicklung relativer Armut über die Zeit nachvollzogen. Eine 'SWOT'-Analyse identifiziert die Stärken, Schwächen, Chancen und Risiken der Genossenschaften. Durch eine Wertschöpfungskettenanalyse werden Informationen über die Akteure, Machtverhältnisse und Informationsflüsse sowie über Produzentenanteile an den Kaffee-Einzelhandelspreisen gewonnen. Um die Erfahrungen der Bauern mit den Kaffee-Zertifizierungssystemen festzuhalten, wird eine thematische Analyse auf qualitative Daten angewandt, die auf der Entwicklung eines individuellen Codierungssystems zur Textreduktion basiert.

In der Forschungsregion sind die Kaffee-Ernteerträge von konventionellem und zertifizierten Kaffeekleinbauern aufgrund unzureichend gepflegter Kaffeeplantagen in der Regel 40% bis 50% niedriger als der nationale Durchschnitt. Höchste Erträge (im Durchschnitt ca. 480kg/ha) werden von den Bio-Bauern erzielt, aber die Kaffee-Erträge variieren, wie bei konventionellen und Bio-Fairtrade zertifizierten Bauern, zwischen den Genossenschaften (von 293kg/ha bis 516kg/ha). Im Vergleich mit konventionellen Kaffeepreisen erhalten Bio-Fairtrade zertifizierte Bauern 11% und biologisch zertifizierte 8% höhere Loco-Hof-Preise; es existieren jedoch Preisunterschiede zwischen den Genossenschaften. Biologische Produktionsprozesse erfordern weniger zugekaufte Produktionsmittel (z.B. Dünger, Spritzmittel), sind aber arbeitsaufwändiger. Aufgrund einer eingeschränkten Verfügbarkeit von Familienarbeitskräften müssen u.a. deswegen zusätzliche Arbeitskräfte eingestellt werden. Diese gleichen die Einsparung bei den Produktionsmitteln aus. Die höheren Loco-Hof-Preise für zertifizierten Kaffee kompensieren die Produktionskosten, führen aber bei den Bio-Fairtrade zertifizierten Bauern im Vergleich zu den konventionellen Bauern nicht zu höheren Deckungsbeiträgen und Gewinnen pro Hektar. Aufgrund ihres höheren Ernteertrags haben Bio-Bauern einen höheren Deckungsbeitrag und Gewinn pro Hektar. Sie haben mit 328US$/ha einen signifikant höheren Profit als die Bio-Fairtrade Bauern mit 147US$/ha und die konventionellen Bauern (191US$/ha). Da sie jedoch in der Tendenz auf kleineren Flächen Kaffee anbauen und mehr Familienmitglieder haben, führt die Steigerung ihrer Deckungsbeiträge im Vergleich zu den anderen beiden Gruppen nicht zu höheren netto pro-Kopf Einkommen aus der Kaffeeproduktion. Auch Bio-Fairtrade zertifizierte Bauern haben keine höheren netto pro-Kopf Einkommen aus der Kaffeeproduktion als konventionelle Bauern.

Ein höherer Anteil der Haushalte biologisch und Bio-Fairtrade zertifizierter Bauern liegt unterhalb der extremen Armutsgrenze als bei den konventionellen Kleinbauern (45% im Vergleich zu 30%), was bedeutet, dass sie nicht ihren Nahrungsmittelbedarf decken können. Zwischen 60% und 70% der konventionellen und der zertifizierten Kaffeebauern haben Einkommen unterhalb der nationalen Armutslinie.Um mehrere Armutsdimensionen und ihre langjährige Entwicklung zu untersuchen, wird eine Hauptkomponentenanalyse angewandt. Dabei wird gezeigt, dass über einen Zeitraum von zehn Jahren biologisch zertifizierte Bauern relativ ärmer geworden sind. Im Jahr 1997 hatten alle Gruppen ein ähnliches relatives Armutsniveau. Die Bio-Fairtrade zertifizierten Bauern konnten ihren relativen Armutsstatus während der Kaffeekrise (in 2002) verbessern; es ging ihnen relativ besser als den konventionellen Bauern. Seitdem sind sie im Vergleich zu den konventionelle Bauern relativ ärmer geworden.

Unabhängig davon, ob die Kaffeebauern zertifiziert sind oder nicht, leiden Nicaraguas Kaffeekleinbauern während zwei bis drei Monaten pro Jahr an Nahrungsmittelknappheit. Während dieser Zeit suchen sie außerhalb ihres Betriebes Arbeit und beantragen informelle und formelle Produktionskredite. In vielen Fällen werden die Produktionskredite für Konsumgüter, hauptsächlich Nahrungsmittel und Medizin, verwendet und nur zum Teil in die Kaffeeplantage investiert. Folglich bleiben die zu erntenden Kaffee-Erträge niedrig, was zu niedrigen Einkommen und erneuten Kreditbedarf führt. Bei unzureichenden Kenntnissen im Management der Betriebsfinanzen oder oder bei Beantragung von mehreren Krediten, die zusammen die Zahlungsfähigkeit uebersteigen, laufen Kleinbauern Gefahr, in einem Teufelskreis der Verschuldung zu landen.

Jede Genossenschaft hat ein einzigartiges Geschäftsmodell, sie unterscheiden sich zum Beispiel in der Mitgliederanzahl, Funktionen und Dienstleistungsangeboten, in der internen Organisation und finanziellen Mitteln. Trotz der unterschiedlichen Geschäftsmodelle der Genossenschaften wählen sie oft die gleichen 'Upgrade-Strategien' wie andere Genossenschaften, hauptsächlich Zertifizierung, Verbesserung der Qualität und eigene Weiterverarbeitung des Kaffees. Eine Analyse der Stärken, Schwächen, Chancen und Bedrohungen (SWOTs) zeigt, dass die Genossenschaften gewisse SWOTs gemeinsam haben, aber auch dass es genossenschafts-spezifische SWOTs gibt. Die gemeinsame Stärke der Genossenschaften ist das Qualitätspotenzial der Region. Die gemeinsamen Schwächen sind der

Mangel an Zugang zu Krediten, ein schwaches Beratersystem und eine schlechte ländliche Infrastruktur. Die gemeinsamen Bedrohungen der Genossenschaften sind der hohe Wettbewerb zwischen nationalen Kaffeekäufern und den Genossenschaften, Korruption und Misswirtschaft, und, den qualitativen Interviews zufolge, zunehmende mikroklimatische Variationen und unregelmäßige Niederschläge. Die gemeinsamen Chancen reichen von vermehrter horizontaler Koordination zur Reduktion der Transaktionskosten über Anteilsurkunden der Mitglieder in der Genossenschaft zu einer verbesserten Transparenz bei den Abrechnungen. Die qualitative Auswertung zeigte keine offensichtliche Verbindung zwischen der Kaffee-Zertifizierungsstrategie der Landwirte bzw. ihrer Genossenschaft und den Kaffee-Deckungsbeiträgen, die die Bauern erwirtschaftet haben. Die 'Upgrade- Strategien' der Genossenschaften, ihre Stärken und Schwächen sowie die Anzahl der auf Kaffee bezogenen Dienstleistungen, die die Genossenschaft ihren Mitgliedern anbietet, scheinen die Kaffee-Deckungsbeiträge stärker zu beeinflussen als die biologische oder die Bio-Fairtrade Zertifizierung.

Die Kleinbauern haben keine Verhandlungsmacht bezüglich der Kaffeepreise – unabhängig von der Wertschöpfungskette an der sie teilnehmen. Zertifizierte Genossenschaften haben eine begrenzte Verhandlungsmacht gegenüber ihren Kunden im Vergleich zu den konventionellen Genossenschaften. In allen Wertschöpfungsketten bestehen Machtasymmetrien zwischen Kaffeekäufern und Verkäufern. Die Quantität und Qualität der Informationsflüsse hängt von der Genossenschaft und der jeweiligen Wertschöpfungskette ab. Informationsasymmetrien sind in der Tendenz in den zertifizierten Ketten geringer, dies hängt jedoch auch von der Genossenschaft ab. Bio-Fairtrade zertifizierte Wertschöpfungsketten neigen im Vergleich zur konventionellen Kette dazu, vor allem in den Kaffe-Konsumländern mehr und kleinere Akteure zu haben. Dies erhöht die Transaktionskosten in den zertifizierten Wertschöpfungsketten und führt damit zu wesentlich geringeren Produzentenanteilen an den Einzelhandelpreisen von Kaffee (8-15% Anteil in zertifizierten im Vergleich zu 24%-34% in konventionellen Wertschöpfungsketten).

Die präsentierten Ergebnisse hängen stark von jeder Genossenschaft ab und es gibt große Unterschiede innerhalb der biologischen und Bio-Fairtrade zertifizierten Genossenschaften. Höhere Loco-Hof-Kaffeepreise führen nicht unbedingt zu einem höheren netto pro-Kopf Kaffee- und Haushaltseinkommen, da die Ernteerträge, Produktionskosten, Familiengröße und Kaffeefläche sowie die Verfügbarkeit von Arbeitskräften eine wichtige Rolle spielen. Die biologische oder Bio-Fairtrade Zertifizierung als 'Upgrade-Strategie' scheint nur dann erfolgreich, wenn das Geschäftsmodell einer Genossenschaft, ihre Stärken, Schwächen und weiteren 'Upgrade-Strategien' unterstützend wirken. Angesichts der oben aufgeführten Punkte ist eine gut funktionierende Genossenschaft für den Erfolg von Zertifizierungen eine notwendige, aber nicht hinreichende Bedingung. Dies zeigt das Beispiel einer gut geführten Bio-Fairtrade zertifizierten Genossenschaft, deren Mitglieder niedrige Kaffee-Deckungsbeiträge haben.

Die wichtigsten Ursachen für anhaltende Armut der Kaffeekleinbauern im Norden Nicaraguas scheinen nicht das Fehlen eines Marktzugangs oder sogenannte 'unfaire' Handelsbedingungen zu sein. Die qualitativen Forschungsergebnisse zeigen, dass Gründe für Armut, u.a., in mangelnden unternehmerischen Fähigkeiten und Managementkompetenzen der Landwirte und Genossenschaften, Finanzdefiziten und Verschuldung von Bauern sowie in der schlechten ländlichen Infrastruktur liegen. Die quantitativen Ergebnisse deuten ferner auf das geringe Kaffee-Ertrags- und Produktivitätsniveau, sowie Land- und Familienarbeitskräftemangel hin. Zertifizierungssysteme befassen sich nicht mit diesen Problemen oder sind nicht in der Lage, diese zu lösen. Die Preise für zertifizierte Kaffees können nicht eine geringe Produktivität oder einen Mangel an Land und Familienarbeitskräften ausgleichen.

Daher können Kaffee-Zertifizierungssysteme nur Teil einer tragfähigen Entwicklungspolitik für arme Kleinbauern in Nicaragua sein; die oben erwähnten produktionstechnischen, infrastrukturellen, organisatorischen und institutionellen Probleme benötigen verstärkte Aufmerksamkeit der politischen Entscheidungsträger. Es wird empfohlen, dass eine Politik, die auf eine Erhöhung der kleinbäuerlichen (Kaffee-) Einkommen zielt, sich auf landwirtschaftliche Produktionsaspekte und den institutionellen Kontext der Kleinbauern und ihrer Genossenschaften konzentriert. Die Kaffee-Erträge müssen gesteigert werden, zum Beispiel durch Investitionen in Forschung für stress-tolerante und lokal-angepasste ertragreiche Kaffeesorten, die den mikroklimatischen Schwankungen Rechnung tragen. Die Kaffeequalität in der Region sollte weiter durch eine entsprechende Kaffee-Sektorstrategie auf nationaler Ebene gestärkt werden, welche ein nationales Kaffeeinstitut oder eine Kaffeeförderation, ähnlich den Modellen in Kolumbien oder Costa Rica, beinhalten sollte. Dies sollte durch Investitionen in die ländliche Infrastruktur begleitet werden. Es wird empfohlen, ein effizientes Beratungssystem zu etablieren, das auch die unternehmerischen Fähigkeiten der Kleinbauern verbessert. Dies könnte auch in Form einer Unterstützung zur Gründung von Beratungsringen geschehen, welche regional arbeiten und sich aus Mitgliedsbeiträgen finanzieren könnten.

Um eine bessere Anbindung der Kleinbauern an (hochwertige) Vermarktungskanäle zu erzielen und das Einkommen der Bauern zu erhöhen, empfiehlt es sich, die Struktur und Funktionsweise der Genossenschaften und ihrer jeweiligen Wertschöpfungsketten zu verbessern. Betriebswirtschaftliche und strategische Beratung für Genossenschaften ist notwendig, da die Genossenschaftsvorsitzenden und das Management nicht adäquat für die internationalen Märkte ausgebildet sind, in denen sie sich behaupten müssen. Ein nationales Banksystem, das auch Genossenschaften Kredite anbietet (zum regulären Marktzinssatz und Marktbedingungen) würde die Abhängigkeit von Exporteuren oder internationalen Kreditgebern vermindern und könnte die Liquiditätsengpässe von Genossenschaften reduzieren. Eine obligatorische jährliche externe Rechnungsprüfung der Genossenschaften, wie sie in anderen Ländern üblich ist, wird als wichtig erachtet, um Missmanagement in Genossenschaften zu verhindern. Diese würde auch die Kreditwürdigkeit der Genossenschaften für die Banken erhöhen.

Die Handels-, Verarbeitungs- und Vermarktungseffizienzen müssen in den biologischen und vor allem in den Fairtrade zertifizierten Wertschöpfungsketten im alternativen, fairen Handelsbereich der Kaffee-Konsumländer verbessert werden. Der alternative Handel ist durch viele kleine Unternehmen oder Non-Profit-Organisationen gekennzeichnet. Die am alternativen, fairen Handel beteiligten Akteure könnten konsolidieren, um Skaleneffekte auszuüben und so ihre Transaktionskosten zu reduzieren. Auch wenn eine Reduktion der Akteure in Kaffee-Konsumländern eine neue Denkweise im alternativen, fairen Handel ist, könnte dies einen wirksamen Beitrag für den Anteil der Landwirte am Einzelhandelspreis leisten und die Loco-Hof Kaffeepreise weiter verbessern.

Resumen Ejecutivo

En los últimos años, el café con certificación orgánica y la doble certificación orgánica y de comercio justo ha crecido en popularidad entre consumidores con una conciencia social, ambiental, y de salud. Dado que los consumidores pagan precios más altos para estos cafés certificados, se supone que, en comparación al café convencional, se pagan mejores precios a los productores y también, que una proporción más alto del valor agregado en países consumidores alcanza a los productores. Por lo tanto se supone que certificaciones de café beneficiarán a los productores de café. El café es un importante producto de exportación para muchos países en desarrollo. La mayoría de la producción mundial de café proviene de alrededor de 20-25 millones familias de pequeños productores en los países en desarrollo. Como certificaciones individuales son demasiado caras los pequeños agricultores tienen que participar en las organizaciones de agricultores, por ejemplo, las cooperativas, con el fin de acceder a la certificación de grupo la cuál es más barato. Los gobiernos y los donantes internacionales apoyan a los esquemas de certificación de café asumiendo que estas certificaciones conectan los agricultores a los mercados de alto valor, aumentan los ingresos de los productores, cambian las asimetrías de poder e información en las cadenas de valor, y contribuyen a la reducción de la pobreza. Sin embargo, solamente existe una evidencia empírica débil que justifique este apoyo. Hay pocos estudios cuantitativos que han aplicado técnicas de muestreo aleatorio, y que analizaron los efectos de los esquemas de certificación en relación de presupuestos parciales, utilidades económicas, distribución de ingresos y niveles de pobreza de los pequeños productores cafeteros certificados. El papel de las cooperativas para el éxito de los esquemas de certificación ha sido desatendido por la investigación. Los estudios disponibles tienen limitaciones metodológicas; por ejemplo, están basados en sólo métodos cualitativos, no incluyen más de una cooperativa o una norma de certificación, o las cooperativas no son elegidas al azar.

Esta investigación trata de llenar las deficiencias identificadas en el conocimiento y la metodología. A través de una combinación de investigación cualitativa y cuantitativa, las estrategias de producción y comercialización de los pequeños productores cafeteros en el norte de Nicaragua están comparados entre los productores que están organizados en cooperativas convencionales, en cooperativas certificadas orgánicas, y cooperativas con la doble certificación orgánica-comercio justo. El análisis comprende (i) el nivel de hogar de los pequeños agricultores y (ii) el nivel organizacional e institucional con respecto a las cooperativas y sus respectivas cadenas de valor de café. El estudio tiene como objetivo, en primer lugar, la identificación de los costos y beneficios socio-económicos de la participación en las cadenas de café orgánico y orgánico-comercio justo, con respecto al nivel de los ingresos del hogar y los ingresos por el café, tal como a la pobreza del hogar. En segundo lugar, se examina cuál es el papel de las organizaciones de agricultores, sus respectivos modelos de negocio y estrategias de mejora (*'upgrading strategies'*), para el éxito o el fracaso de los esquemas de certificación. En tercer lugar, la integración de los caficultores y sus cooperativas en la cadena de valor del café, la estructura y el funcionamiento de las cadenas de valor y el efecto de valor agregado de la certificación es examinada.

El estudio se llevó a cabo en el norte de Nicaragua en los departamentos Madriz, Nueva Segovia y Matagalpa en las fincas cafeteras situadas entre 900 y 1300 msnm. El café de todos los agricultores se clasificó como 'estrictamente de altura' (*'Strictly High Grown'*); la especie de café es *Coffea Arabica*. El diseño de la muestra aseguró que la región de la investigación fue homogénea con respecto a la condiciones de vida, nivel socio-económico, y así como a las las características de crecimiento del café que dirigen el rendimiento de los productores cafeteros. Después de haber seleccionado las cooperativas al azar, se seleccionó de estas cooperativas 327 hogares cafeteros al azar y se les entrevistó con un cuestionario estructurado. La colección de datos cualitativa se realizó con un total de 58 entrevistas con personales claves, 67 entrevistas semi-estructuradas de agricultores y 24 grupos de discusión con agricultores cafeteros. Los datos primarios estaban colectados durante dos estancias de investigación en 2007 y 2008.

La investigación analiza los presupuestos parciales, las utilidades contables y las utilidades económicas de la producción de café. Se mide el ingreso del hogar y se elabora un índice de incidencia de la pobreza. Se usa un análisis de componentes principales para determinar los niveles actuales de pobreza relativa y el desarrollo de la pobreza relativa en el tiempo. Un análisis FODA identifica las fortalezas, oportunidades, debilidades, y amenazas de las cooperativas. A través de un análisis de la cadena de valor, se obtiene la información sobre los actores, flujos de información, distribución del poder y de los precios. Para la identificación de las experiencias de los agricultores con los esquemas de certificación de café, se aplica un análisis temático a los datos cualitativos mediante el desarrollo de un sistema de código individual para la reducción de datos.

En la región de la investigación, los rendimientos de café de pequeños productores convencionales y certificados son por lo general de 40% a 50% inferior que el promedio nacional, debido a las actividades de mantenimiento limitadas y plantaciones de café mal gestionadas. Los rendimientos más altos (en promedio alrededor de 480kg/ha) se obtienen por los productores orgánicos, pero los niveles de rendimiento varían, igual como de los productores convencionales y certificados orgánicos-comercio justo, entre las cooperativas (abarcando desde 293kg/ha a 516kg/ha). Los cafés certificados orgánicos y orgánicos-comercio justo lograron en promedio un 8% y 11% respectivamente por encima de los precios convencionales; diferencias de precios entre las cooperativas también existen. Los procesos de la producción orgánica requieren menos insumos comprados, pero son más laboriosos. Debido a la disponibilidad limitada de mano de obra familiar, más labor tiene que estar contratado, lo cual contrarresta los gastos de producción ahorrados. Los precios más altos de cafés certificados compensan por los costos de producción pero fallan en incrementar los presupuestos parciales por hectárea y las utilidades económicas en el caso de agricultores certificados orgánicos-comercio justo en comparación con productores convencionales. Debido a los rendimientos más altos, los productores orgánicos obtienen un aumento de los presupuestos parciales por hectárea y las utilidades económicas. Con 328US\$/ha tienen una utilidad económica significativamente mayor que la de los agricultores certificados orgánico-comercio justo (147US\$/ha) y los agricultores convencionales (191US\$/ha). Sin embargo, como suelen tener áreas cafeteros más pequeños y familias más grandes, el incremento en el presupuesto parcial no se traduce en mejores ingresos de café per cápita para los productores certificados orgánicos comparado a los otros dos grupos. Los productores certificados orgánicos-comercio justo, tampoco tienen mejores ingresos de café per cápita que los productores convencionales.

Entre los productores certificados orgánicos y orgánicos-comercio justo, una mayor proporción de los hogares se agrupa debajo de la línea de la pobreza extrema que entre los productores convencionales (45% comparado con el 30%) – lo cual significa que los productores orgánicos y orgánicos-comercio justo no pueden cubrir sus necesidades alimenticias. Entre el 60% y el 70% de los productores convencionales y certificados están por debajo de la línea de pobreza nacional. Usando el análisis de componentes principales para investigar varias dimensiones de la pobreza y su desarrollo con el tiempo, se encontró que durante un período de diez años, los productores certificados orgánicos se hicieron relativamente más pobre. En el año 1997, todos los grupos tenían niveles similares de pobreza relativa. Los productores certificados orgánicos-comercio justo primero mejoraron su situación de pobreza relativa durante la crisis del café (en 2002) y estaban relativamente mejor económicamente que los productores convencionales. Desde entonces, el nivel de pobreza relativa de los productores orgánicos-comercio justo se deterioró en comparación con los productores convencionales.

Independientemente de si los productores recibieron la certificación o no, los pequeños agricultores cafeteros enfrentan dos o tres meses de escasez de alimentos por año, durante la cual buscan empleo fuera de la finca y solicitan créditos formales e informales. En muchos de los casos, se usa el crédito para las necesidades inmediatas de consumo, como comida o medicinas y sólo parcialmente se invierte en el cafetal. En consecuencia, los rendimientos de cosecha se mantienen bajos, lo que tiene como consecuencia ingresos bajos y nuevos necesidades de crédito. Cuando los agricultores son financieramente analfabetos o solicitan créditos más altos que su capacidad de pago, es probable que entren en un círculo vicioso del endeudamiento.

Cada cooperativa tiene un modelo de negocio único; difieren, por ejemplo en la cantidad de miembros, las funciones y servicios, organización interna, y las características financieras. A pesar de sus diferentes modelos de negocio, las cooperativas frecuentemente suelen elegir las mismas estrategias de mejoramiento como de las otras cooperativas – principalmente certificación, calidad y procesamiento propio. El análisis de fortalezas, oportunidades, debilidades y amenazas (FODA) mostró que las cooperativas tienen ciertos FODAs en común pero también hay FODAs específicos por cooperativa. La fortaleza en común de las cooperativas es el potencial de calidad de la región. Las debilidades en común están relacionadas con la falta de acceso a crédito, y con un sistema débil de extensiónismo, y una infraestructura rural deficiente. Las amenazas comunes de las cooperativas son la alta competencia entre compradores nacionales de café y las cooperativas, corrupción y mala administración, y según las entrevistas cualitativas, el aumento en la variación micro-climática y los patrones de lluvias poco confiables. Las oportunidades comunes oscilan entre más coordinación horizontal para reducir los costos de transacción, compartir certificados reconociendo las posesiones de miembros en la cooperativa, e incrementar la transparencia sobre las deducciones en los pagos. La evaluación cualitativa indicó que no había una asociación obvia entre la estrategia de certificación de café de los agricultores/su cooperativa y los presupuestos parciales de café que obtuvieron los agricultores. Las estrategias de mejoramiento de las cooperativas, las fortalezas y debilidades, junto con la cantidad de servicios relacionados con el café que las cooperativas ofrecen a los productores, suelen ser más relacionados a los presupuestos parciales de café que la certificación orgánica u orgánica-comercio justo.

Se ha revelado que los agricultores no tienen poder de negociación sobre los precios independientemente de la cadena de valor, mientras que las cooperativas certificadas tienen un poder limitado de negociación frente a sus compradores en comparación con las cooperativas en la cadena convencional. El poder está distribuido desigualmente entre compradores y vendedores de café en todas las cadenas. La cantidad y calidad de los flujos de información depende de la cooperativa y el modelo de la cadena de valor. Las asimetrías de información son menores en las cadenas certificadas; sin embargo, esto también depende de la cooperativa. Las cadenas de valor certificadas orgánicas-comercio justo suelen tener más actores y actores de menor tamaño, especialmente en países consumidores, en comparación con la cadena convencional. Esto incrementa los costos de transacción en las cadenas de valor certificadas y por lo tanto lleva a disminuir de manera importante la proporción de los productores al precio final de la venta de café (8% - 15% en cadenas certificadas comparado a 24% - 34% en cadenas convencionales).

Los resultados presentados dependen mucho de cada cooperativa y hay grandes variaciones dentro de las cooperativas certificadas orgánicas y orgánicas-comercio justo. Se puede concluir que el aumento de precios pagado al productor no necesariamente llega a aumentar los ingresos netos de café por cápita o los ingresos del hogar, debido a que los niveles de rendimiento, costos de producción, el tamaño de familia y del terreno, así como la disponibilidad de mano de obra juegan un papel importante.

La certificación orgánica y orgánica-comercio justo, como una estrategia de mejoramiento, sólo parece exitosa cuando el modelo de negocio de una cooperativa, sus fortalezas, debilidades y otras estrategias de mejoramiento son respaldante. Dado a las limitaciones mencionadas anteriormente, una cooperativa altamente funcional es una condición necesaria pero no suficiente. Esto fue demostrado por el ejemplo de una cooperativa certificada orgánica-comercio justo que funcionó bien pero cuales productores tenían presupuestos parciales bajos.

Las principales causas de la persistencia de la pobreza entre los pequeños productores
cafeteros en el norte de Nicaragua no parecen ser la falta de acceso a los mercados ni las
condiciones comerciales llamadas 'injustas'. Con base en el análisis cualitativo, las razones
de la pobreza son falta de habilidades empresariales y de gestión de los agricultores y del
personal de las cooperativas, el analfabetismo financiero y el endeudamiento de los agricultores, así como la infraestructura rural muy débil. Con base en los resultados cuantitativos, posibles razones para la pobreza son los bajos rendimientos, los bajos niveles de
productividad, las limitaciones de terreno y de mano de obra. Los esquemas de certificación no pueden abordar ni resolver estos problemas. Los precios de café certificado no
pueden compensar por la baja productividad ni por las limitaciones de terreno o mano
de obra.

Por lo tanto, los esquemas de certificación solamente pueden ser parte de una política de
desarrollo viable para los pequeños agricultores pobres en el norte de Nicaragua; los problemas de producción, de infraestructura, los problemas organizacionales e institucionales
mencionados arriba requieren más atención de los responsables políticos. Se recomienda
que las políticas, las cuales tienen como objetivo aumentar los ingresos de pequeños productores de café a través de estrategias de mejoramiento, deberían enfocarse no sólo en
los aspectos de producción, sino también en el contexto institucional de los pequeños agricultores y sus cooperativas. En cuanto a la producción de café, las políticas deberían
enfocarse a los niveles de rendimiento de café, por ejemplo, mediante inversiones en investigación en variedades mejoradas, tolerantes a factores adversos, y adaptadas localmente
para enfrentar las variaciones micro-climáticas. La calidad del café en la región debería
estar fortalecida con una estrategia de apoyo del sector cafetero al nivel nacional, la cual
debería incluir un instituto o federación nacional de café tal como existe en Colombia o
Costa Rica. Esto debería estar acompañado de inversiones en infraestructura rural. Se
recomienda establecer un sistema de extensión eficiente que también promueve las capa-
cidades empresariales de los agricultores. Esto podría ser también en forma de facilitar
el establecimiento de asociaciones de extensión, las cuales podrían operar regionalmente
y estar financiados por medio de las contribuciones de sus propios miembros.

Para poder vincular mejor los agricultores a los mercados (de alto valor) y aumentar
sus ingresos, se recomienda enfocarse más en la estructura y funcionamiento de las organizaciones de productores y sus respectivas cadenas de valor. Asesoramiento estratégico
y comercial para las cooperativas es necesario, como los dirigentes y el personal de las
cooperativas no están preparados para los mercados internacionales en las cuales tienen
que participar. Un sistema bancario ofreciendo créditos a las cooperativas (con tasas de
interés y condiciones de préstamo de mercado), reduciría la dependencia de los exportadores o proveedores internacionales de crédito y podría aliviar limitaciones de liquidez de
las cooperativas. Se considera importante de introducir una auditoría externa de las cooperativas la cuál debería ser obligatoria y anual, como existe en otros países, para reducir

la mala gestión de una cooperativa. Esta auditoría, también, aumentaría la solvencia de las cooperativas para los bancos.

Las eficiencias de comercio, procesamiento, y de la comercialización en las cadenas de valor de la certificación orgánica, pero especialmente de la doble certificación orgánica-comercio justo, tienen que mejorarse en el sector de comercio alternativo con su gran cantidad de empresas y organizaciones pequeñas de lucro o sin fines lucros. Estos actores podrían consolidar para ejercer las economías de escala y reducir sus costos de transacción. La consolidación es sin duda una nueva forma de pensar en el sector del comercio alternativo, pero efectivamente podría contribuir a mejorar las proporciones las cuales reciben los agricultores del precio final de venta en los países consumidores y aumentar el precio de café que recibe el caficultor.

Acknowledgements

During the course of this research, which included field trips to three countries in Latin America, many people crossed my path – some accompanied me from the beginning, others came at the end; some walked along my side for a long, others for a short while, some only for a brief moment. Yet, each of these encounters left its traces on my path – some made the research more enjoyable, more interesting, stimulating and challenging or more rewarding – others made it more complicated and more questionable – while again a third group just showed me the world through their eyes and thus contributed indirectly to my research.

I would like to thank Prof. Dr. Manfred Zeller for supervising my PhD research and for the opportunity to be a research associate at his chair of 'Rural Development Theory and Policy' at the University of Hohenheim. I am very grateful for his support, advice and valuable comments throughout the course of my research. I am also very thankful to Dr. Thomas Oberthür for his interest in this research, for connecting me to the *International Center for Tropical Agriculture (CIAT)* and his research project on high-value crops, for his support and the inspiring discussions. Further thanks go to Prof. Dr. Anne Bellows and Prof. Dr. Stephan Dabbert for being the co-reviewer and additional examiner, respectively. I greatly appreciate the financial support for my field research through the *Small Grant Fund* of the *Advisory Service on Agricultural Research for Development (Beratungsgruppe für Entwicklungsangewandte Agrarforschung – BEAF)* of the former *Deutsche Gesellschaft für Technische Zusammenarbeit (GTZ)*, now *Deutsche Gesellschaft für Internationale Zusammenarbeit (GIZ)*.

I am very grateful to Dr. Axel Schmidt, then the regional director of CIAT in Nicaragua and his assistant Vilia Escóbar for their manyfold support, especially with logistical issues. Being with my thanks already in Nicaragua, I would like to extend my thanks to Gonzalo Castillo for his invaluable support and excellent introduction into Nicaragua's culture and coffee sector, to Heberto Rivas who became a great source of coffee knowledge and friend, to Silvio Leite, head judge of the CoE, and all the other professional coffee cuppers and roasters of the CoE who passed the spirit of coffee on to me, to Evelyn López and Clark Davies, who tracked coffee farmers even in the most remote areas and supported data capturing, to Dr. Danilo Saavedra for his personal and his organization's interest in the research topic, to Santos, Juana and their children who shared their life, trapped in extreme poverty, with me like I was one of them, and finally to Dr. Rein van der Hoek and his wife Jannie who offered a lovely place to stay when being in Managua. Since a field research without willing and contributing farmers is not possible – I am very thankful to all who participated and shared their time and experience.

In Germany, Colombia and elsewhere, my many thanks go to Anna Kiemen, Rick Peyser and Don Seville for their insights in coffee value chains, to Dr. Nicole Schönle-

Acknowledgements

ber for our (not only) scientific discussions, to Gudrun Contag, Katharina Mayer and Coni Schumacher and their helping hands in any administrative matter, their smile and encouragement, to Dr. Detlef Virchow, Teodardo Calles, Peter Laderach, Carolina Gónzalez, Jenny Correa, Herman Usma. My many thanks also go to my colleagues from the institute, especially from the chair of Rural Development Theory and Policy, for all the valuable discussions, critical reflections, and the sharing of not only research experiences but of ideas, cultures and many special moments.

Finally, I would like to say my deep thanks to my mother, Susanne Beuchelt, for her continuous support and help in innumerable ways and for her belief in me, and to Lutz Göhring. No words of thanks can acknowledge his contribution in the last years: For convincing me that 'Latex' is great and for helping out in those moments where it prove to be as unpredictable as 'Word', for his love, his help, the wonderful discussions and critical questions about my research.

I only named some persons who contributed to the research while many more were leaving their traces on my path. To those not mentioned I still wish to express my deep gratitude as the experience to have crossed their paths was more rewarding than any treasure.

Chapter 1

Introduction

This research work analyzes the participation and welfare effects of coffee certification on smallholders in northern Nicaragua. In addition, the role of the smallholders' cooperatives and their respective value chains are investigated in order to identify other factors which may contribute to the success or failure of coffee certification schemes.

1.1. The coffee sector in brief

Coffee is mainly produced in developing countries by approximately 25 million of farmers, the majority of whom are smallholders (Gresser and Tickel, 2002). In Central and South America 70% of coffee farmers operate less than five hectares of land. Also in Nicaragua, the production is dominated by smallholders and involves around 20% to 40% of the rural labor force (Lewin et al., 2004). Nicaragua's export earnings from coffee amounted to 26% in the year 2000, yet decreased to 14% in 2002. Since then, the share recovered and varied between 15% and 20%. Coffee is a product with a high price volatility; frequent crises are common (Cashin et al., 2002). Exported coffee experienced a continuous price decline in the past decades. From 1977 to 2001, real international prices for coffee fell by 5.1% yearly (Cashin et al., 2002). In 2000, the real coffee price was around 50% of the price of the mid-1960s and a mere 20% of the peak market values in 1977 (Fitter and Kaplinsky, 2001). The reason for the break-down in Nicaragua's coffee export earnings in 2002 was the most severe coffee crisis of the past 100 years which lasted from 1998-2003 (Lewin et al., 2004). In many regions coffee prices were temporarily falling below the production costs (Lewin et al., 2004; Raynolds et al., 2004). This coffee crisis reduced producers' income strongly; and especially smallholders were severely hit by the price decline (Raynolds et al., 2004). Raynolds et al. (2004) indicate that thousands of producers in Central America abandoned their coffee parcels since conventional prices did not even cover harvesting costs during the coffee crisis. Between 1998 and 2001, poverty rates of Nicaraguan smallholder coffee farmers increased by 2% while the poverty rate among rural households dropped by 6% due to economic growth in the rest of the country (Lewin et al., 2004). Working and living conditions deteriorated, education and health became unaffordable, child labor increased, farmers and laborers migrated to the cities or to other countries. As farmers cut costs, the reduced use of inputs allowed pests to flourish, which led to declining coffee yields (Varangis et al., 2003).

At the same time while coffee producers were struggling to survive, coffee was booming in the coffee consuming industrialized countries. This phenomenon came to be known as the 'coffee paradox' (Daviron and Ponte, 2005). In the last decade, coffee became a fashionable drink and coffee bar chains, like Starbucks, have expanded rapidly (ibid.). While sales in the conventional coffee sector have been stagnant, new consumption patterns have emerged and the so-called differentiated coffee segment has experienced strong growth (Lewin, Giovannucci, and Varangis, 2004). Differentiated coffees can be distinguished by those that emphasize quality aspects (such as gourmet and specialty coffees) and those that stand out for 'sustainability' aspects such as a specific production technology or trading system (e.g., organic, shade-grown or fair trade coffees). They account for around 9%-12% of the traded volume and a larger percentage of the profits in industrialized countries (ibid.). During the past few years, the sustainable coffee segment has grown around 20%-25% annually (Giovannucci et al., 2010). This calls the interest of the coffee industry, as differentiated coffee allows competitive positioning in high-value markets (Henson and Reardon, 2005). In these market segments consumers pay price premiums which lie considerably above prices paid for conventional coffee (Giovannucci et al., 2010).

Among consumers, three types of buying patterns for differentiated coffee can be identified. The first type buys gourmet quality coffee. The second type buys 'good' coffee yet is not focusing on and paying for the material quality but mostly for symbolic quality and in-person service[1] in coffee bar chains which sell an ambience and a certain social positioning (Ponte, 2002). The third buying pattern is shown by consumers to whom food safety, environmental and social attributes of food products and production processes have become increasingly important (Asfaw et al., 2010; Basu et al., 2003; Codron et al., 2006; Ponte and Gibbon, 2005; Raynolds, 2000; Rigby and Cáceres, 2001; Swinnen and Maertens, 2007). This interest was triggered by food safety crises and workers' rights scandals from the mid-1990s onwards which brought environmental and ethical issues to the attention of a broad range of consumers (Codron et al., 2006). While consumption of organic and fair trade certified products in industrialized countries started in the 1970s, these products were only offered locally and represented a small market niche. Since 2000, food products which are produced and processed according to environmental standards, like organic farming practices, have moved from niche market existence to mass markets. With the introduction of fair trade labeled products in large supermarket chains and discounters, a similar move to mass markets is likely to occur for these ethical and fair trade products (Codron et al., 2006).

---

[1]Coffee quality can be distinguished between material quality, which refers to the material attributes of a product which are objective and measurable, symbolic quality and in-person service quality. Symbolic quality refers to attributes that cannot be easily measured by human senses or technological devices. Symbolic quality attributes are based on reputation and can refer to certifications such as organic or Fairtrade certification. In-person-service quality refers to attributes in regard of interpersonal relations, i.e. the relation between the person delivering the service and the consumer and/or the relations between the consumers. In person-service takes place at the point of consumption which can be cafés or coffee chains such as 'Starbucks' (Daviron and Ponte, 2005).

In an attempt to identify ways out of the recent coffee crisis which had struck the coffee farmers, differentiated coffee market niches are considered by farmers, cooperatives, policy-makers, NGOs, and donors as a promising alternative to conventional coffee channels (Kilian et al., 2006; Lewin et al., 2004; Linton, 2008; Willer and Yussefi, 2007). Roasters and retailers are also interested in differentiated coffee markets as they are motivated by profits, market shares and social responsibility.

Therefore, smallholder coffee producers and their producer organizations are supported to obtain sustainable coffee certifications like the organic, Fairtrade[2], 'Shade-grown' or 'Bird-friendly', 'Rainforest Alliance' or 'UTZ Certified' certification. There are two private company standards that also became important: 'Starbucks C.A.F.E. Practices' and 'Nespresso' which is a brand belonging to 'Nestlé'. Furthermore, there is the Common Code for the Coffee Community (4C-Codex) which is verification-based but not certified by an independent assessment. The organic and the Fairtrade certification are two of the oldest and best-known certifications[3]. The other certifications are less common among smallholders; yet the coffee volumes traded of other certification schemes have strongly increased in recent years (Giovannucci et al., 2010). Together, all the sustainability initiatives made up more than 8% of all green coffee exports in 2009 (ibid.). Environmental and social aspects are important in organic production and in the Fairtrade system. However, the focus of the two systems is different. While organic production lays its emphasis on environmental aspects, Fairtrade emphasizes social standards, trading and marketing relationships. Guided by similar principles, Fairtrade also aims at increasing the share of organic production under their label. The double certification Organic-Fairtrade becomes very popular among coffee buyers and consumers. Both, Fairtrade and organic certification claim to contribute to poverty reduction and food security in developing countries (IFOAM, 2006a,b; Wills, 2006). Certified coffee markets have experienced strong growth in Europe (Murray and Raynolds, 2006) and continue to grow in the USA where growth rates of 56% for organic and 33% for Fairtrade coffee are reported (Giovannucci and Villalobos, 2007). Fairtrade coffee consumption world wide has been growing annually between 11% and 20% in the past five years (FLO, 2007; Giovannucci et al., 2010).

In this context, the importance of marketing cooperatives has often been highlighted as a link between consumers and producers that allows farmers to participate in new market developments and high-value chains (Bacon, 2005; Varangis et al., 2003; Wollni et al.,

---

[2]The term 'fair trade' is written in lower case and two words where we refer to the movement of trading goods fairly. When the term 'Fairtrade' is capitalized we refer to one specific fair trade standard and label, the 'Fairtrade certification' of the 'Fairtrade Labeling Organizations International' (FLO). Apart from the Fairtrade standard of FLO, there exist other fair trade standards developed by other certification agencies like the 'Fair for life' standard of the international 'Institute for Marketecology' (IMO). Yet, during the time of research, these other fair trade standards were not very common in smallholder coffee production.

[3]They are described in detail in the following chapter.

2010). The coffee value chain is split into two parts: the coffee producing countries, mostly developing countries with millions of smallholders involved, and consuming countries, mainly industrialized countries with comparatively rich consumers (Ponte, 2002). In coffee producing countries, the producers' value captured for conventional coffee production and primary processing over the past 30 years has varied between 10% and 20% of the final retail value (Daviron and Ponte, 2005; Fitter and Kaplinsky, 2001; Talbot, 1997). In the 1970s and 1980s, the total value captured in consuming countries was between 40% and 50%; by the end of the 1980s this share had risen to over 70% (Daviron and Ponte, 2005; Talbot, 1997).

The direct trading relationships in Fairtrade chains should lead to a higher share of the retail price being returned to the producer than in the conventional chain (Arnould et al., 2009). According to Slob (2006, p 134) the *"percentage of the final retail value that is retained by the producers is much higher in the Fair Trade system than in the mainstream market"*. Certified coffees are marketed at a higher retail price than conventional coffees. This price premium is generally assumed to compensate for higher production costs and/or lower yields inherent to organic farming and to give a 'fair' price covering production and living costs in the Fairtrade sector (Wills, 2006). Therefore, Fairtrade is seen as a *"feasible alternative to the unfair distribution of value that characterizes today's mainstream coffee market"* Slob (2006, p 139). Organic movements also claim benefits to producers and rural development through the enhancement of economic, social and environmental sustainability (IFOAM, 2006a,b). Yet as Fitter and Kaplinsky (2001) indicate, it is not necessarily given that the returns to differentiation are captured by producers even when they can meet the requirements of the differentiated specialty markets.

1.2. Problem statement

While the coffee prices have recovered in recent years and begin to lead to a new peak since winter 2010, several questions remain: Whether the higher coffee prices at retail level are actually passed on to farmers, how much benefit farmers derive from the added value of certification schemes in comparison to other chain actors, how certification affects their income and their poverty level, and which role the cooperatives as marketing institutions are playing for the success of certification schemes (Giovannucci et al., 2010). Due to regulation, cooperatively organized farmers are the only producers of Fairtrade coffee. Also organic certified coffee is often produced by cooperatives or associations since certification is too costly for individual smallholders (Rice, 2001).

The effects of organic and Fairtrade certification on coffee prices have been more frequently investigated than those on income and poverty. Bacon (2005) found out that during the coffee crisis participation in organic and Fairtrade networks lead to higher coffee prices, reduced farmers' livelihood vulnerability and the risk to loose land titles. Fairtrade certified cooperatives in Mexico and Central America paid their members prices that were two to three times higher than those of local middle-men (Raynolds et al., 2004). A study in Costa Rica also shows that participation in specialty coffee marketing channels as well as in cooperatives serves to increase farm-gate prices (Wollni and Zeller, 2007). Other researchers observe that the higher prices from certification lead for example to better nutrition, increased education, health, improved household sanitation systems, water supplies or cooking stoves (Becchetti and Costantino, 2008; Raynolds et al., 2004; Utting-Chamorro, 2005; Arnould et al., 2009; Bacon et al., 2008). Fairtrade and organic coffee certification also increase social organization and contribute to capacity building of farmers and their organizations (Bray et al., 2002; Raynolds et al., 2004; Taylor et al., 2005).

Yet, research results on certified coffee prices are not always that clear-cut positive. Analysing data from several countries in Central America, Kilian et al. (2006) show small to large price differentials for organic and Fairtrade coffee for the 2002/03 harvest. In some occasions, Fairtrade farm-gate prices were much lower than conventional prices, for example when existing cooperative debts needed to be paid (Bacon, 2005; Utting-Chamorro, 2005). Despite participating in certified markets, farmers still reported a decline in their quality of life during the coffee crisis as income from coffee sales to alternative markets was not enough to offset additional difficulties farmers faced (Bacon, 2005). According to Arnould et al. (2009) and Valkila (2009), the positive farm-gate price differences of organic and Fairtrade coffee continue after the crisis in Guatemala, Nicaragua, and Peru while Philpott et al. (2007) do not locate premiums in Mexico.

Most studies emphasize the higher prices paid in certified market channels and deduce that higher prices lead to higher farm income which then reduces poverty. As net coffee income is not only determined by the price but also by yield levels and production costs, this conclusion is premature. Certified coffees have distinct production and marketing systems, thus associated costs differ from those of the conventional system. In addition, the rising quality standards for organic and Fairtrade coffee (Murray and Raynolds, 2006; Raynolds, 2009) increase production costs as more labor is needed. There is not much information about production costs of organic certification schemes (Kilian et al., 2006). Kilian et al. (2006) and Van der Vossen (2005) find that organic farmers have higher production costs than conventional farmers who are not compensated adequately by price premiums for organic or Fairtrade coffee. Other studies mention lower production costs (Nigh, 1997; Valkila, 2009). Mutersbaugh (2002) indicates that in Mexico organic coffee production is only successful when farmers have high yields and that premiums, at best, just cover production costs (Mutersbaugh, 2005). In contrary, Bray et al. (2002) show for Mexico that higher prices for organic coffee offset higher production costs and that farmers benefit from participation.

Qualitative research on the past ten years of coffee certification schemes as well as research combining qualitative with some quantitative data, yet without random sampling and statistic analysis, is now abundant – for example Bacon et al. (2008); González and Nigh (2005); Kilian et al. (2006); Murray and Raynolds (2006); Mutersbaugh (2002, 2005); Raynolds et al. (2004); Utting-Chamorro (2005); Valkila (2009). There are very few quantitative studies evaluating organic or Fairtrade certification schemes (Arnould et al., 2009; Bacon, 2005; Becchetti and Costantino, 2008; Bolwig et al., 2009) based on a random sample with a treatment and control group. Similar findings are stated by Becchetti and Costantino (2008) for evaluations of the Fairtrade certification in general.

Additionally, there are not many scientific studies that verify the claims of organic production and fair trade certification systems regarding poverty alleviation (Murray and Raynolds, 2006). The analysis of certified coffee value chains as compared to conventional value chains has been addressed by only two studies both using data from the coffee crisis (Daviron and Ponte, 2005; Mendoza and Bastiaensen, 2003). Mendoza and Bastiaensen (2003) indicated not much higher producer shares of final retail prices for certified coffees than for conventional ones, while Daviron and Ponte (2005) found higher shares for Fairtrade but not for organic certified coffee farmers.

Not only prices and shares of the final retail price the coffee producers receive should be questioned, but also how power is distributed and what information is communicated in the value chain. Both are determining factors for the successful integration of smallholder producers into the value chain (Gereffi et al., 2005; Talbot, 2002). Power and information issues have been overlooked in research on certified coffee value chains (Tallontire, 2009). The above mentioned studies only marginally touch the business model of farmer organizations and their upgrading strategies apart from certification. The fit of upgrading strategies to the business model of a farmer organization as well as the determinants for the success of upgrading through certification has so far been neglected by research. This may also be one explanation for the contradictory findings in respect to the economic benefits in the above cited studies.

Summarizing, no study so far has systematically investigated production costs, profitability, household income or the poverty status of organic and Fairtrade certified coffee producing households based on a quantitative analysis with randomly selected producers and triangulated that with qualitative data. In addition, most of the cited studies on certification effects have neither randomly sampled cooperatives/farmer organizations nor analyzed several cooperatives or certification standards simultaneously.

At present, only weak empirical evidence is available to international donors and governments as a basis for their support to coffee certification schemes as tools which link farmers to high-value markets, increase producers' incomes and shares of final retail price, which change the power and information relations between buyers and sellers, and which contribute to poverty reduction. However, for national governments and international donors to make effective policies for poverty reduction, quantitative data about the profitability and poverty reduction potentials of certified coffee production as well as about the factors leading to the success of certification schemes are of great importance.

1.3. Research objectives and questions

This paper contributes to filling the identified knowledge gaps through an analytical comparison of conventional, organic and Organic-Fairtrade certified small-scale coffee farmers, cooperatives, and value chains. The analysis of participation and socio-economic effects in certified coffee supply chains on small-scale farmers requires approaches from different viewpoints for a holistic presentation and understanding of the functioning and linkages between the different production systems, market channels, and chain actors, since the coffee value chain is very complex. This research is based on two analytical levels: (i) the household level of the smallholder coffee producers and (ii) the organizational and institutional level regarding the cooperatives and value chain. The study first aims to identify the socio-economic costs and benefits of participation in organic and Organic-Fairtrade certified coffee chains with respect to level of coffee and household incomes as well as household poverty. Second, the role of the producer groups, their business models and upgrading strategies for the success of certification schemes is explored. Third, the integration of coffee farmers and their cooperatives into the coffee value chain, the structure and functioning of the value chains and the value adding effect of certification is examined. Quantitative data are triangulated with qualitative data.

The following research questions are analyzed:

1. Regarding the small-scale coffee farming household level
   a) Which types of smallholders participate in organic and Organic-Fairtrade certified coffee marketing channels in comparison to conventional marketing channels?
   b) What are the smallholders' costs and benefits of participation in organic and Organic-Fairtrade certified coffee chains with respect to level of household income and poverty?
   c) How do organic and Organic-Fairtrade coffee certifications relate to the production, food security and financial situation of smallholders in Nicaragua?
2. Regarding the organizational and institutional level of conventional and certified[4] cooperatives
   a) What are the business models and upgrading strategies of conventional, organic and Organic-Fairtrade certified cooperatives and how well do they work?
   b) How are the coffee gross margins linked to the organic and Organic-Fairtrade certification strategy, to the cooperative's business model and its other upgrading strategies?
3. Regarding the organizational and institutional level of the conventional and certified coffee value chains

---

[4]For simplification, the term 'certified' is used when speaking about both the organic and Organic-Fairtrade certification.

a) What are the structures of the Fairtrade and Organic-Fairtrade certified chains compared to a conventional coffee value chain in terms of actors, governance, and information flows?

b) What is the effect of adding value to coffee through Fairtrade and Organic-Fairtrade certification on farm-gate prices and the producers' share of the final retail price?

1.4. Outline of the dissertation

The dissertation is presented as a cumulative thesis. The first part comprises Chapter 2, which provides a short overview of the organic and Fairtrade certification standards and the conceptual framework, and Chapter 3 which describes the research area and the sampling design applied in this research. The second part addresses the research questions in form of articles submitted to peer-reviewed journals. Figure 1.1 provides an overview of the linkages between the chapters. Chapter 4 and 5 address the household level of coffee farmers. Chapter 4 contains the first paper published in *Ecological Economics* and describes the conventional and certified coffee smallholders, their coffee gross margins and profits, household income and their relative as well as absolute poverty level. Chapter 5 provides a predominately qualitative analysis of the food security status as well as of production and financial constraints of conventional and certified farmers. With the help of seasonal calendars, it analyzes work load, food, income and expenditure patterns and discovers a vicious cycle of indebtedness. It was submitted in February 2011 to the *Quarterly Journal of International Agriculture*.

Chapter 6 and 7 address the organizational and institutional level. Chapter 6 embraces a paper accepted by the journal *Renewable Agriculture and Food Systems*. It discusses the role of cooperatives, their business models and upgrading strategies as well as the strengths, weaknesses, opportunities, and threats of cooperatives and links these to the farmers' coffee gross margins. Chapter 7 contains a paper submitted in January 2011 to the journal *Agriculture and Human Values*. It compares the conventional with organic and Organic-Fairtrade certified value chains and analyzes the value adding effects of certification schemes, especially regarding farmers' shares of retail prices.

Chapter 8 provides a concluding discussion and synthesis of the obtained results and gives an outlook on desirable follow up research as well as policy implications. This is followed by the references as far as they did not appear at the end of each article and the appendices.

Figure 1.1.: The structure of the dissertation

Source: Own illustration.

Part I.

Conceptual Framework, Research Area and Design

Chapter 2

The Certification Standards and the Conceptual Framework

This chapter first describes the organic and Fairtrade certification standard. Then the conceptual framework is outlined.

2.1. The organic and Fairtrade certification standard

Organic coffee production has to follow national standards either defined by the producing countries, if existent, and/or by the importing countries. Additionally, specific standards of certain organic associations, e.g. 'Demeter' or 'Naturland', need to be followed when the label of these associations is to appear on the product. These additional standards may be requested by the coffee buyers. There is a worldwide umbrella organization for the organic movement called 'International Federation of Organic Agriculture Movements' (IFOAM). It is uniting more than 750 member organizations in 108 countries and also provides organic standards. An internationally accepted definition of organic agriculture in the sense of an ISO-standard does not yet exist. IFAOM has approved in March 2008 a definition of organic agriculture which, given the large diversity of IFOAM's members, comes close to an international definition (IFOAM, 2009a): *"Organic agriculture is a production system that sustains the health of soils, ecosystems and people. It relies on ecological processes, biodiversity and cycles adapted to local conditions, rather than the use of inputs with adverse effects. Organic agriculture combines tradition, innovation and science to benefit the shared environment and promote fair relationships and a good quality of life for all involved."*

In 2005, the General Assembly of IFOAM adopted the 'Principles of Organic Agriculture' which are seen as guide for organic agriculture development, programs and standards. There are four principles on which organic agriculture is based: The principle of health, of ecology, of fairness and of care (IFOAM, 2009b). These principles affect and alter the livelihoods of farmers. The principle of health involves that organic agriculture should sustain and enhance the health of soil, plants, animals and humans. This includes the prohibition of synthetic fertilizers, herbicides and pesticides as well as the use of certain animal drugs or food additives. It encourages the use of organic fertilizer and other inputs based on the natural available resources. But there is more to organic agriculture than just a change of the input system. Following the principle of ecology, organic agriculture should be based on and integrated in the living ecological systems and cycles and protect them. Social aspects are included in the principle of fairness. It is stated that those involved in organic agriculture should build on human relationships that ensure fairness at all levels and to all parties – farmers, workers, processors, distributors, traders, and consumers. IFOAM (2009b) states that: *"Organic agriculture should provide everyone involved with a good quality of life, and contribute to food sovereignty and reduction of poverty. [...] Fairness requires systems of production, distribution and trade that are open and equitable and account for real environmental and social costs"* . The last principle, the principle of care, demands that organic agriculture should be managed in a precautionary and responsible manner to protect the health and well-being of current and future generations and the environment.

IFOAM emphasizes that organic agriculture is seen increasingly as a major contributor to poverty reduction and food security in developing countries (IFOAM, 2006a,b). Hence, based on a holistic production management system, organic agriculture is said to contribute to rural development through enhancing governance, creation of vibrant economic spaces through offering employment opportunities, maintenance of a healthy environment and building social capital in rural areas (IFOAM, 2006b).

The World Fair Trade Association (WFTO) sees fair trade as *"a response to the failure of conventional trade to deliver sustainable livelihoods and development opportunities to people in the poorest countries of the world"* (WFTO, 2010). WFTO defines fair trade as *"a trading partnership, based on dialogue, transparency and respect, that seeks greater equity in international trade. It contributes to sustainable development by offering better trading conditions to, and securing the rights of, marginalized producers and workers - especially in the South"* (WFTO, 2010). In 1997 the 'Fair Trade Labelling Organization' (FLO) was created to coordinate and harmonize the various fair trade labeling initiatives (Wills, 2006). Like organic agriculture, the FLO's Fairtrade standard also follows several key principles. The first key principle claims Fairtrade to be a strategy for poverty alleviation and sustainable development through creating opportunities for economically disadvantaged producers or for producers that are marginalized by the conventional trading system. Fairtrade builds on transparency and accountability; moreover it is seen as a means to develop producer's independence through capacity building. The presumably most known principle is the payment of a 'fair' price. A 'fair' price is defined as a price in the regional or local context that has been agreed on through dialogue and participation, covering not only the costs of production but also of living and enabling production which is socially just and environmentally sound (Wills, 2006). 'Fair' traders should ensure prompt payment and, if possible, assist with or provide pre-harvest or pre-production financing. Gender equity refers to the proper valuation of and reward for women's work as well as to female empowerment in the organizations involved with Fairtrade. Safe and healthy working conditions and the encouragement for better environmental practices are also among the key principles (Wills, 2006).

In addition to the general standards and principles, there are specific standards for each Fairtrade product. For coffee, FLO standards require that producers must be small, family-based growers organized into politically independent democratic associations which preserve the natural resources (Murray and Raynolds, 2006). The Fairtrade standards cover the two species of coffee which are mainly traded: *Coffea Arabica* (Arabica coffee) and *Coffea Canephora* (Robusta coffee). Arabica coffee is generally considered to be of higher quality and better taste. It is mainly grown in Latin America, in Central and East Africa, in India and to some extent in Indonesia. Robusta coffee is mainly grown in Central and West Africa, in South-East Asia and to some extent in Brazil (FLO, 2008).

The Fairtrade pricing scheme is special. In coffee, it involves a minimum price below which the coffee price is not allowed to fall. The minimum price refers to the 'Free on board' (FOB) price which is usually paid to the cooperatives. From this FOB coffee price, the processing, exporting costs (including loading the ship) and any other costs occurring at the cooperative are deducted, so that the farm-gate coffee price is lower than the Fairtrade minimum price. When the world-market price is higher than the Fairtrade minimum price, the reference market price applies. For Arabica coffee, this is the price based on the New York "C" contract of the IntercontinentalExchange (ICE), the former New York Board of Trade (NYBOT). A Fairtrade premium is paid on top of the market price or the minimum price of conventional coffee (FLO, 2008). This premium is to be used for social projects which contribute to the social development of the community and the coffee farming households.[1]

The Arabica Fairtrade minimum prices and premiums paid to cooperatives in 2007 and 2008 are listed in Table 2.1. Prices are distinguished between washed and non-washed Arabica coffee, which refers to the processing method after harvest. In Nicaragua, all coffee is washed and then dried, and thus classifies as washed Arabica. Non-washed Arabica coffee is produced through drying the coffee berry with the surrounding pulp, followed by removing the dried pulp from the berry. The two processing methods differ in the labor requirements and lead to different tastes of the coffee. In addition to the Fairtrade minimum price and the Fairtrade premium, organic certified producers receive an additional premium. The Fairtrade differential for organic certified coffee and the fairtrade premiums have been increased in 2007[2]. From June 2008 onwards, also the Fairtrade minimum prices have been raised and the regional discrimination eliminated. One reason for the price increase were complaints from coffee producers and their cooperatives, especially from Latin America, over the fact that the minimum prices were too low in relation to the production costs.

In Nicaragua, coffee was the first organic product produced and sold internationally. By the end of the 1980s, the first 200 certified coffee bags were sold to a German buyer, the

---

[1]The development of the Arabica coffee prices and the relationship between conventional and Fairtrade coffee prices is presented well in a graph by the Fairtrade Foundation (2011) for the years from 1989 until 2010.

[2]The price increase took place after the coffee harvest and the coffee sales had been finished for the coffee year 2006/07, for which the production and marketing data had been collected in this research. Thus, the price increase did not affect the collected data.

Table 2.1.: Fairtrade minimum prices and premiums for Arabica coffee in US\$/lb, for 2007 and 2008

| Year | Type of coffee | Fairtrade minimum price (for conventional coffee) | | Organic differential | Fairtrade premium |
		Central America, Mexico, Africa, Asia	South America, Carribean Area	All regions	All regions
Until 31st May 2007	Washed Arabica	1.21	1.19	0.15	0.05
Until 31st May 2007	Non-washed Arabica	1.15	1.15	0.15	0.05
June 1st 2007 - May 31st 2008	Washed Arabica	1.21	1.19	0.20	0.10
June 1st 2007 - May 31st 2008	Non-washed Arabica	1.15	1.15	0.20	0.10
From June 1st 2008 onwards	Washed Arabica	1.25	1.25	0.20	0.10
From June 1st 2008 onwards	Non-washed Arabica	1.20	1.20	0.20	0.10

Source: FLO (2007, 2008).

'One-World-Shop' called 'Dritte-Welt Partner'. In 1986, another solidarity coffee buyer in Germany was founded, the enterprise 'Mittelamerika und Export Kaffee' (MITKA) with the objective to import coffee from Nicaragua, to support Nicaragua's socialist government, and to inform German citizens about the political situation, especially the civil war, in Nicaragua. The first organic coffee trade with Germany was thus based on solidarity, and rather politically motivated than by the organic certification (Garibay and Zamora, 2003). Thus, Germany has a long trading relationship of certified coffees with Nicaragua. In 2007/08, Nicaragua was the fifth largest producer of organic coffee (Figure 2.1) and the third most important exporter of Fairtrade certified coffee (FLO, 2010).

2.2. The conceptual framework

The presented research is based on two conceptual frameworks taking into account the two dimensions of research. The first framework focuses on the household level while the second framework concentrates on the organizational and institutional level. The frameworks build the underlying concepts for the submitted research articles.

The conceptual framework for the household level was developed departing from the sustainable livelihoods approach of Scoones (1998). There are five core asset categories or

Figure 2.1.: Organic green and roasted coffee by origin for 2007/08 in 60-kg bags

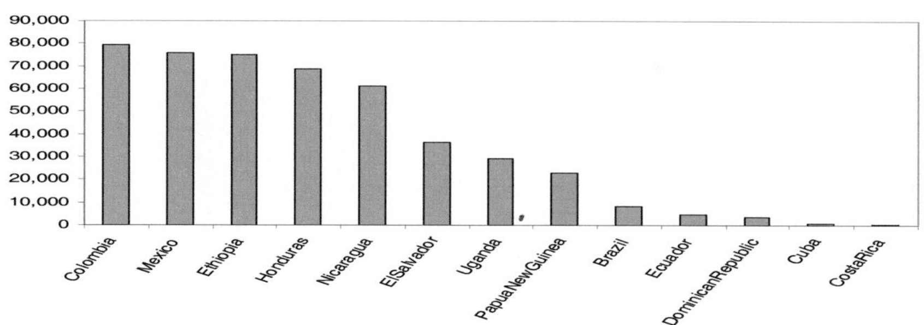

Source: According to data from ICO (2009).

types of capital upon which livelihoods are based: human, natural, financial, social, and physical capital (Figure 2.2). These assets determine the access to cooperatives, to factor and product markets as well as to formal and informal institutions. The decision of a farmer to participate in cooperatives and in coffee certification is assumed to depend on the utility a farmer attributes to participation. Where there are substantial benefits to be obtained through collective action, households will get involved (Varughese and Ostrom, 2001).

The factor and product markets, institutions, and cooperatives (with their assets and structures) influence the livelihood assets of small-scale coffee producers and the farm households' performance and transaction costs. They form the basis of the producers' production and certification strategy. When people embark on a production and certification strategy, the five types of capital are the resources they can use and combine in order to achieve a certain outcome. The production and certification strategies are hypothesized to influence positively the production and marketing performance of coffee producers through higher and more stable prices, access to credits and extension for coffee production. This leads to certain outcomes such as increased income, improved food security and living conditions or better farm management. When positive, the outcome of the chosen strategies contributes to reduce household vulnerability to stress such as volatile coffee prices and unreliable microclimatic conditions. The success or failure of the production and marketing strategies influence positively or negatively the livelihood assets and the vulnerability of farmers.

The conceptual framework addressing the organizational and institutional level is presented in Figure 2.3. Summarizing, it builds on the coffee producing members, cooperatives, and the coffee value chains. The vulnerability context is the same as in the conceptual

Figure 2.2.: Factors determining the participation and welfare effects of coffee certifications on small-scale farmers

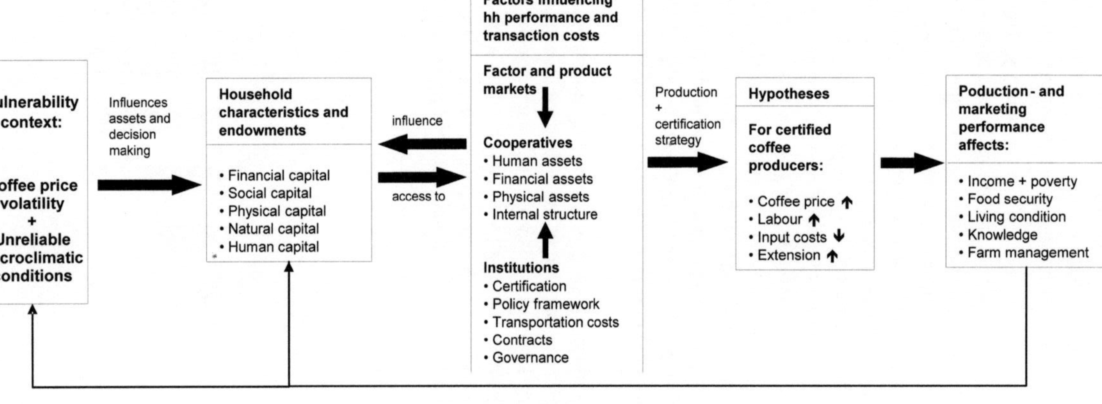

Source: Own illustration.

Figure 2.3.: Cooperative and value chain characteristics which influence coffee producing
cooperative members and the success of certification schemes

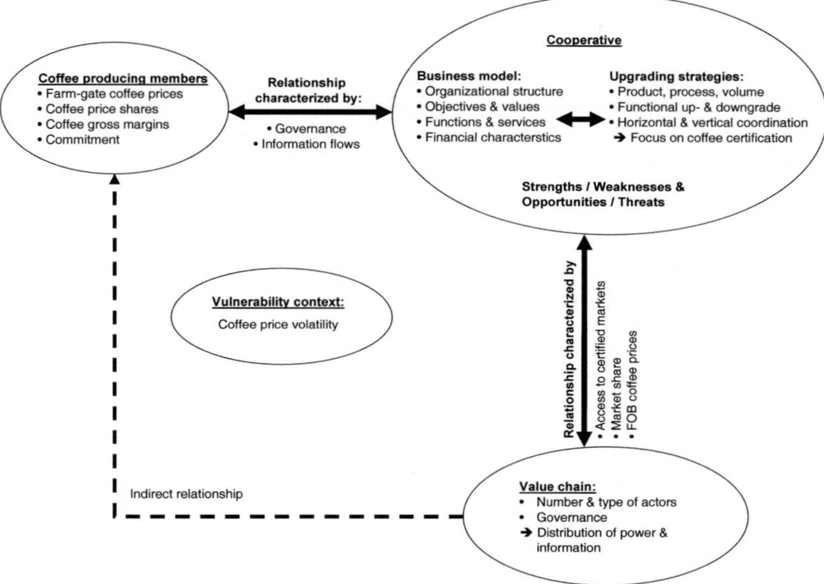

Source: Own illustration.

framework for the household level; dominated by coffee price volatilities and unreliable microclimatic variations. The cooperative can be described by its business model, its chosen upgrading strategies of which the coffee certification strategy is one, and its strengths, weaknesses, opportunities, and threats. The term business model refers to organizational structure, size, values, functions and services, and financial characteristics. Upgrading strategies can be in the field of product, process or volume upgrades, functional up- and downgrades and horizontal and vertical coordination. The coffee value chain is characterized by the number and type of actors in it, the type of governance (i.e. the organization of who does what; determined by the power relationships between actors), and the distribution of information and power. These characteristics determine part of the transaction costs and the coffee prices. The relationship between the cooperative, which is also a value chain actor, and the other actors in the value chain is characterized by the access to certified markets, the share of coffee sold to these markets and the FOB coffee prices.

The analysis of value chain actors is limited on the cooperative's coffee buyers and the subsequent value chain in industrialized countries. Other potential value chain actors like input suppliers or credit providers have not been the target of this research.

The characteristics of the value chain affect the farm-gate coffee prices the cooperative can offer and thus also the coffee gross margins possibilities of the cooperative's members. The farmer's commitment to the cooperative depends i.a. on the coffee prices, the cooperative business model and upgrading strategies. The cooperative is also influenced by its members' commitment, their loyalty and reliability regarding coffee delivery and credit payment. The relationship between the cooperative and its members can be characterized by the governance, the distribution of power and information flows. Governance in global value chains can be understood as the organization of activities within the chain; the who does what in a value chain is determined by the power relationships between actors. Some actors have the power to define rules and standards, to include or exclude chain participants, or to decide upon the value-adding activities of each participant. There is an indirect relationship between the other value chain actors and the coffee producers via the coffee prices and the farmer's price share of the final retail prices.

— Chapter 3 —

Research Area and Research Design

This chapter first describes the research area and provides selected background information for the research region. Then the research design with the sampling framework and
questionnaires are explained.

3.1. Research area

Nicaragua has around 43,200 coffee farms covering a total of around 130,800 hectares. Of
these farms, 52% have a maximum size of seven hectares – the limit until which they
are still considered as smallholdings in Nicaragua. The survey was conducted in the
northern departments Madriz and Nueva Segovia on coffee farms situated between $900m$
and $1,300m$ a.s.l. For the value chain analysis, also coffee producers of the department
of Matagalpa were interviewed. The departments Madriz and Nueva Segovia produce together 13% of Nicaragua's coffee on 20% of the coffee area, the Matagalpa region produces
around 33% of the coffee on 24% of the area (IICA, 2004). An overview of the Arabica
coffee production in the four major production departments and national average is given
in Table 3.1[1]. Based on field observations, the much lower level of coffee yields in Nueva
Segovia and Madriz is assumed to relate to lower levels of technification and smaller farms;
data to confirm this statement are not available.

These four departments are known for their good coffee quality. The coffee of all
farmers in the sample was classified as strictly high grown, which achieves higher prices
on the world market than other classifications; the species is *Coffea Arabica*. The sample
design ensured that the research region was homogeneous with respect to living conditions,
socio-economic level and coffee growing characteristics.

Poverty is widespread in Nicaragua and especially in the research region. At country
level, 46% of the population are poor. Like in many poor developing countries, poverty
is largely a rural phenomenon: of the poor, 29% live in urban areas and 68% in rural
areas (World Bank, 2008). The departments of Madriz, Nueva Segovia and Matagalpa
are all classified with high levels of poverty. In Madriz, around 37% of the population are
extremely poor, in Nueva Segovia 34% and in Matagalpa 32%; the majority of the poor
in all departments is again concentrated in rural areas (Gobierno de Nicaragua, 2001).

---

[1]To compare with, Wintgens (2009) classifies yield levels around 300kg/ha as low, 600kg/ha as
medium, 1,200kg/ha as high and above 2,000kg/ha as very high.

Table 3.1.: Coffee area, production and yield in the four major production departments of Nicaragua and national average, 2000/01

Department	Area (%)	Production (%)	Yield in kg/ha
Jinotega	33.3	45.4	1040.1
Matagalpa	24.4	33.1	1033.5
Nueva Segovia	11.9	7.1	458.7
Madriz	8.4	5.7	518.1
Total country	100.0	100.0	762.8

Source: IICA (2004).

3.2. Research design and sampling framework

Data were collected during two field trips in 2007 and 2008. Quantitative household data were collected in 2007 by random sampling, complemented by qualitative data with most interviews undertaken in 2008.

For the household survey, a two stage random-sampling procedure was applied. In the first step, the cooperatives were randomly selected; in the second step the cooperative's coffee producing members were randomly selected. As no complete list of existing cooperatives was available, a cooperative list was constructed for the research region during several visits of the region. The cooperatives were classified according to their certification and market channel in conventional, organic and Organic-Fairtrade certified cooperatives. Cooperatives had to be certified for a minimum of five years. From this list, half of the cooperatives were randomly selected in each market channel, which resulted in seven cooperatives with one having conventional as well as organic producers. Depending on the cooperative, either a random sampling or a two-stage random sampling was applied to select participating members. 327 cooperative members were surveyed with a structured questionnaire. Coffee producing members were classified, like the cooperatives, into three groups according to their production: conventional, organic and Organic-Fairtrade. In those cases where a household had several coffee producers, all were interviewed and the data aggregated at the household level. Data of one conventional cooperative could not be integrated as it turned out at the end of the data collection period that the list with cooperative members was based only on members which had registered themselves in an extension program and another list did not exist. This lowered the sample size of conventional farmers.

The household questionnaire (Appendix A) included household demographic characteristics including the employment and income types; data on dwelling-related indicators such as housing size, material, cooking fuels, access to electricity, water and telephones; data on food-related indicators such as recent consumption patterns and hunger periods; and other asset-based indicators such as transport assets, appliances and electronics, farm equipment, livestock, and land ownership. Some data was collected on agricultural production but the focus was on coffee production data such as inputs, expenditures, time requirements, harvest and marketing channels, and coffee prices. Moreover, data were enquired on the certification, membership in the cooperative as well as in other organizations. The questionnaire ended with a brief overview of access to financial services and credits. The quantitative interviews have been conducted by trained local interviewers who had experience with working and interviewing farmers.

For the qualitative interviews, the same cooperatives were chosen. With cooperative staff, exporters, roasters, and researchers 48 key-person interviews were conducted, 33 semi-structured interviews were done with farmers and 21 focus group discussions with farmers and an average group size of nine participants. For each cooperative, the aim was to have one seasonal calendar elaborated by a focus group, one group discussion about the impacts of certification or livelihood changes of conventional farmers, and one group focusing on gender issues. Farmers were selected according to gender, certification, distance to the cooperative headquarters, and their functions in the cooperative. The standardized questionnaire for the cooperatives is attached in Appendix B, the guiding questions of the semi-structured interviews in Appendix C, and the topics for the focus group sessions in Appendix D. In the farmer interviews and groups sessions not all topics/questions were pursued but were handled flexible according to the farmer's story, background and experiences. The guiding questions for the semi-structured interviews of key-persons were individually developed, depending on the professional background and area of expertise of each key-person. These guiding questions are available upon request. All qualitative interviews were conducted by the author. The qualitative interviews have been recorded with the consent of the respondents and have been transcribed afterwards. Additional field notes were taken.

For the value chain analysis, additional semi-structured interviews were conducted with 34 small-scale coffee producers in 22 communities, seven presidents of first-level cooperatives and five cooperative staff members. Also, three focus group workshops with groups consisting of an average of ten farmers in each group were held. In these workshops, the local value chains have been further investigated. Information was gathered on prices, information flows, position of actors in the chain, and types of relationships between actors. The roasters and importers were asked to complete a semi-structured questionnaire which was sent by e-mail after a personal first contact. This questionnaire is also available upon request.

Data were analyzed with the statistical programs SPSS and STATA. Where applicable, the computer-assisted analysis of qualitative data (CAQDAS) has been chosen, the CAQDAS software was called MAXQDA. For identifying the farmers' experiences with coffee certification schemes, a thematic analysis was applied to the qualitative data by developing an individual code system for data-reduction. Due to the cumulate structure of this dissertation, the applied analytical methods and models, e.g. how gross margins, absolute or relative poverty levels were calculated, are described in each paper dealing with the respective topic in the following chapters.

Part II.

Articles, Results and Discussion

Profits and Poverty: Certification's Troubled Link for Nicaragua's Organic and Fairtrade Coffee Producers

TINA BEUCHELT AND MANFRED ZELLER

This chapter has been accepted in January 2011 for publication as article in the journal Ecological Economics.[1]

Abstract

Governments, donors and NGOs have promoted environmental and social certification schemes for coffee producers as certified market channels are assumed to offer higher prices and better incomes. Additionally, it is presumed that these certifications contribute to poverty reduction of smallholders. Yet, gross margins, profits and poverty levels of certified smallholder coffee producers have not yet been quantitatively analysed applying random sampling techniques. Our quantitative household survey of 327 randomly selected members of conventional, organic and organic-fairtrade certified cooperatives in Nicaragua is complemented by over hundred qualitative in-depth interviews. The results show that although farm-gate prices of certified coffees are higher than of conventional coffees, the profitability of certified coffee production and its subsequent effect on poverty levels is not clear-cut. Per capita net coffee incomes are insufficient to cover basic needs of all coffee producing households. Certified producers are more often found below the absolute poverty line than conventional producers. Over a period of ten years, our analysis shows that organic and organic-fairtrade farmers have become poorer relative to conventional producers. We conclude that coffee yields, profitability and efficiency need to be increased, as prices for certified coffee cannot compensate for low productivity, land or labour constraints.

Keywords:

Cooperative; smallholder; gross margin; income; poverty; profitability

---

[1]This is the accepted author manuscript for publication in *Ecological Economics*. A definite version was subsequently published in: Beuchelt, T.D. and M. Zeller (2011). Profits and poverty: Certification's troubled link for Nicaragua's organic and fairtrade coffee producers. Ecological Economics, Volume 70, Issue 7, Pages 1316-1324. DOI: 10.1016/j.ecolecon.2011.01.005. Copyright 2011 Elsevier B.V. This chapter is printed with permission from Elsevier.

4.1. Introduction

Coffee is an important export product for Nicaragua as it contributes 24% to total exports earnings. The production is dominated by smallholders and involves around 20-40% of the rural labour force (Lewin et al., 2004). The last worldwide coffee crisis from 1998/99-2002/03 affected producers' income strongly and in many regions coffee prices were falling below the production costs (Fitter and Kaplinsky, 2001; Raynolds et al., 2004). Smallholders have been among the hardest hit by this price decline. Between 1998 and 2001, poverty rates of Nicaraguan smallholder coffee farmers increased by 2% while the poverty rate among rural households dropped by 6% due to economic growth in the rest of Nicaragua (Lewin et al., 2004). Paradoxically, at the same time, the coffee market in importing countries flourished. Driven by growing social, environmental and health consciousness (Basu et al., 2003), certified coffees have become popular among roasters and consumers in industrialized countries (Daviron and Ponte, 2005; Rice, 2001). Currently, the organic and the fairtrade certification are two of the oldest and most well-known coffee certifications in the market albeit many more exist.

Since the coffee crisis, national governments, NGOs and international donors have promoted the marketing of coffee through group-based, certified market channels as a viable business model for poor smallholders (Kilian et al., 2006; Linton, 2008; Willer and Yussefi, 2007). The changes in consumer demand and policy thinking have led to a growing body of literature investigating the effects of environmental and social certification standards on environmental indicators like tree, mammal, bird, and butterfly species (Gobbi, 2000; Gordon et al., 2007; Philpott et al., 2007) as well as on socio-economic indicators. Before discussing the various studies that focus on socio-economic indicators of organic and fairtrade coffee certification schemes, the following briefly describes these two standards. Due to regulation, cooperatively organized farmers are the only producers of fairtrade coffee. Also organic certified coffee is often produced by cooperatives or associations since certification is too costly for individual smallholders (Rice, 2001). The exact standards for organic coffee depend on the importing country and certification label. All standards focus on enhancing the health of soils, plants, animals, and humans and prohibit the use of synthetic agro-chemical inputs (IFOAM, 2009a,b).

Different labels exist also in the fair trade movement. The most common standard is the 'fairtrade' label which follows several key principles including creating opportunities for economically disadvantaged producers; payment of a fair price; pre-financing; transparency and accountability; capacity building; respect for the environment, and gender equity (Wills, 2006). In reference to coffee, the term 'fair price' is the guaranteed minimum price including a social premium which covers production and living costs[2] (Slob,

---

[2]In 2007, the fairtrade minimum price for conventional washed Arabica coffee in Central America was 2.67US\$/kg coffee 'Free on Board' (FOB), the organic differential 0.33US\$/kg, and the social premium 0.11US\$/kg. From July 2008 onwards, the fairtrade minimum price,

2006). Guided by similar principles, fairtrade also aims at increasing the share of organic production under their label. The double certification organic-fairtrade becomes very popular among coffee buyers and consumers. Both, fairtrade and organic certification claim to contribute to poverty reduction and food security in developing countries (IFOAM, 2006a,b; Wills, 2006).

Several studies conducted during the coffee crisis supported the promotion of certification schemes and showed that organic and fairtrade coffee markets tend to offer higher prices than the conventional market (Bacon, 2005; Daviron and Ponte, 2005; Lewin et al., 2004; Utting-Chamorro, 2005). During the crisis, fairtrade farmers in Mexico, Guatemala and El Salvador had received prices two to three times higher than conventional farmers (Raynolds et al., 2004). Wollni and Zeller (2007) show that in Costa Rica participation in marketing cooperatives and in speciality coffee market channels (which include the certified market channels) lead to higher producer prices. Other researchers observe that the higher prices from certification lead for example to better nutrition, increased education, improved household sanitation systems, water supplies or cooking stoves (Becchetti and Costantino, 2008; Raynolds et al., 2004; Utting-Chamorro, 2005). Fairtrade and organic coffee certification also increase social organization and contribute to capacity building of farmers and their organizations (Bray et al., 2002; Raynolds et al., 2004; Taylor et al., 2005). Bacon (2005) finds that while certified and conventional farmers both reported a decline in their quality of life during the coffee crisis, higher prices for certified coffee had some positive effects, for example a lower fear of loosing the land. Organic and fairtrade certification have also moderate positive effects on education, health and infrastructure investments (Arnould et al., 2009; Bacon et al., 2008).

Yet, research results on prices and thus income are not always that clear-cut positive. Analysing data from several countries in Central America, Kilian et al. (2006) show small to large price differentials for organic and fairtrade coffee for the 2002/03 harvest. In some occasions, fairtrade farm-gate prices were much lower than conventional prices, for example when existing cooperative debts needed to be paid (Bacon, 2005; Utting-Chamorro, 2005). According to Arnould et al. (2009) and Valkila (2009), the positive farm-gate price differences of organic and fairtrade coffee continue after the crisis in Guatemala, Nicaragua, and Peru while Philpott et al. (2007) do not locate premiums in Mexico.

Most studies emphasize the higher prices paid in certified market channels and deduce that higher prices lead to higher farm income which then reduces poverty. As net coffee income is not only determined by the price but also by yield levels and production costs, in our opinion this conclusion is premature. Certified coffees have distinct production and marketing systems with different associated costs than the conventional system. In addition, the rising quality standards for organic and fairtrade coffee (Murray and Raynolds, 2006; Raynolds, 2009) increase production costs as more labour is needed. We

---

organic differential, and premium were raised.

agree with Kilian et al. (2006) that there is not much information about production costs of organic certification schemes. Kilian et al. (2006) and Van der Vossen (2005) find that organic farmers have higher production costs than conventional farmers which are not compensated adequately by premiums of organic or fairtrade prices. Other studies mention lower production costs (Nigh, 1997; Valkila, 2009). Mutersbaugh (2002) indicates that in Mexico organic coffee production is only successful when farmers have high yields and that premiums, at best, just cover production costs (Mutersbaugh, 2005). In contrary, Bray et al. (2002) show for Mexico that higher prices for organic coffee offset higher production costs and that farmers benefit from participation. The cited studies are based on the analysis of non-random samples of coffee farmers which may contribute to the contradictory results.

Qualitative research on the past ten years of coffee certification schemes as well as research combining qualitative with quantitative data, yet without random sampling and statistic analysis, is now abundant – for example Bacon et al. (2008); González and Nigh (2005); Kilian et al. (2006); Murray and Raynolds (2006); Mutersbaugh (2002, 2005); Raynolds et al. (2004); Utting-Chamorro (2005); Valkila (2009). We found very few quantitative studies evaluating organic or fairtrade certification schemes (Arnould et al., 2009; Bacon, 2005; Becchetti and Costantino, 2008; Bolwig et al., 2009) based on a random sample with a treatment and control group. Similar findings are stated by Becchetti and Costantino (2008) for evaluations of the fairtrade certification in general. Arnould et al. (2009) only focus on the benefits of fairtrade certification in regard to coffee revenue, education, and health in three countries in Latin America. Bacon (2005) proves farm-gate price differences for organic and fairtrade coffee producers during the coffee crisis. Bolwig et al. (2009) find positive effects for participation in a certified contract scheme and modest effects for the application of organic farm techniques. Becchetti and Costantino (2008) focus on effects of fairtrade vegetable and fruits in Kenya on crop diversity, market price, price satisfaction, living conditions, food consumption, dietary quality, and child mortality.

Poverty in small-scale farming systems is strongly linked to crop and animal production, therefore it is important to additionally analyse production costs and prices of non-certified farm products as well as total household income. Summarizing, no study so far has systematically investigated production costs, profitability, household income and poverty of organic and fairtrade certified coffee producing households based on a quantitative analysis with randomly selected producers. However, for national governments and international donor agencies to make effective policies for poverty reduction, quantitative data about profitability, poverty development and poverty reduction potentials of certification schemes are of great importance.

This paper contributes to filling this gap with a random sample quantitative study comparing two certification schemes – the organic certification and the double certification organic-fairtrade – against a control group of conventional coffee producers in Nicaragua.

Our study shall provide quantitative evidence to policy makers and donors who are currently supporting or planning to support certification schemes as a tool to reduce poverty. We analyse the following three research questions:

1. Which types of smallholders participate in conventional, organic and organic-fairtrade certified coffee production schemes?

2. In comparison to conventional producers, what are the income effects of participation in certified[3] coffee production?

3. How did poverty of conventional and certified small-scale producers evolve in the past ten years and are there any differences in their current poverty level?

Our hypotheses are:

Ad 1. The organic coffee production system is more laborious than the conventional system but labour requirements are covered by available family labour. As organic coffee production requires less expenditures for purchased inputs we thus expect lower variable production costs than in the conventional coffee production.

Ad 2. (a) Prices at farm-gate are higher in certified than in conventional market channels. (b) These higher prices are sufficient to cover additional costs of participation and lead to net increases in per hectare and per capita coffee income as compared to conventional producers.

Ad 3. Ten years ago conventional and certified farmers had equal poverty levels. Since then, relative poverty levels of households participating in organic and organic-fairtrade developed more positively than the conventional producers.

Additionally, for hypotheses (2) and (3) we expect organic-fairtrade producers to fare even better than the organic producers given fairtrade certification's stronger focus on poverty reduction and rural development.

The next section describes the data collection process and methods used for measuring coffee profitability, income and poverty. This is followed by the results and discussion before conclusions regarding the profitability and poverty effects of certification schemes are drawn.

4.2. Methodology

In this section, we first describe the survey area and data collection process. Then we explain the methods used for measuring profitability as well as absolute and relative poverty.

---

[3]For simplification, we use the term 'certified' when speaking about both organic and organic-fairtrade certified coffee.

4.2.1. Data collection

The survey was conducted in the northern Nicaragua departments Madriz and Nueva Segovia on coffee farms situated between 900m and 1300m a.s.l. The coffee of all farmers was classified as strictly high grown; the species is *Coffea Arabica*. The sample design ensured that the research region was homogeneous with respect to living conditions, socio-economic level, as well as coffee growing characteristics driving performance of coffee farmers. Quantitative household data were collected in 2007, complemented by qualitative data from 48 key-person interviews with cooperative staff, exporters, roasters and researchers, 33 semi-structured producer interviews and 21 focus group discussions (with an average group size of nine participants) undertaken in 2007 and 2008.

As no complete list of existing cooperatives was available, we constructed a cooperative list for the research region. We classified cooperatives according to their certification and market channel in conventional, organic and organic-fairtrade certified cooperatives. Cooperatives had to be certified for a minimum of five years. From this list, we randomly selected half of the cooperatives in each market channel, which resulted in seven cooperatives with one having conventional as well as organic producers. Depending on the cooperative, either a random sampling or a two-stage random sampling was applied to select participating members. 327 cooperative members were surveyed with a structured questionnaire. Coffee producing members were classified into three groups according to their production: conventional, organic and organic-fairtrade. In those cases where a household had several coffee producers, all were interviewed and the data aggregated at the household level. A conventional cooperative dropped out at the end of the data collection period, which lowered the sample size of conventional farmers.

4.2.2. Measuring profitability

The profitability of smallholder coffee production is measured by gross margins, profits and break-even analyses. The achieved coffee prices are discounted according to the time difference between main month of coffee delivery and final settlement of the bill. This approach is chosen as certified producers receive the final payment long after harvest has finished, at earliest around four to five months, but usually around eight to nine months, after first harvest delivery. The price for certified coffee already includes the direct certification costs the cooperative had to bear. It also includes the direct costs of the internal control system, which is a requirement for organic group certification.

At harvest time, both conventional and certified farmers need additional cash to cover the high costs of harvesting. Therefore, they sell part of their harvest in the spot market at conventional coffee prices. Five different spot market channels are used: intermediaries coming to the farm, intermediaries in villages, exporters, the own or other cooperatives buying coffee at spot market prices. Thus, the total coffee revenue is based on various prices. It is given by

$$R_i = x_i p_i^{disc} + \sum_{j=1}^{5} y_{ij} p_{ij}^{spot} \tag{4.1}$$

where R_i = total revenue from coffee production; i = household index ($i = 1, \ldots, N$); x_i = coffee quantity sold to cooperative; p_i^{disc} = discounted coffee price (where applicable); y_{ij} = coffee quantity sold to spot market channels with $j = 1, \ldots, 5$; and p_{ij}^{spot} = spot market coffee price for the respective market channel. Eq. (4.1) is then used to derive the average price P_i^{aver} a farmer received across all channels, given the equation:

$$P_i^{aver} = \frac{R_i}{x_i + \sum_{j=1}^{5} y_{ij}} \tag{4.2}$$

Gross margins are useful to compare the profitability of different crops or production systems. The equation for gross margin per hectare coffee is given by

$$GM_i = \frac{1}{a_i} \left(R_i - (c_i^{inp} + c_i^{harv} + c_i^{hlb}) \right) \tag{4.3}$$

where GM_i = gross margin of one hectare coffee; R_i = revenue from coffee production (eq. (4.1)); c_i^{inp} = variable input costs (chemical and organic fertilization as well as pest, disease, weed control; nursery and transport costs); c_i^{harv} = variable harvest costs including hired labour for coffee picking (as payment is per quantity not per hour); c_i^{hlb} = hired labour costs; a_i = coffee area in hectare; and i = household index ($i = 1, \ldots, N$). The accounting profit per hectare A_i is the difference between the gross margin (eq. (4.3)) and the fixed costs. The main fixed cost in small-scale coffee production is the depreciation cost of the depulper[4] per ha c_i^{dep}. The accounting profit A_i is thus given by:

$$A_i = GM_i - c_i^{dep} \tag{4.4}$$

This is the return to family labour and management used in coffee production and constitutes the cash income per hectare coffee available to the household. The cash income for the whole coffee area is later compared to the income threshold of the poverty line (see section 4.2.3).

Although not included in the accounting profit of a coffee farm, opportunity costs exist and are important for managerial and policy decisions (Kay et al., 2004) as they indicate the differences in efficiency of resource use in the three production systems. Therefore, we estimate the economic profit through including the opportunity costs for the production factors (Torre, 2002). The economic profit is calculated as

$$E_i = A_i - c_i^{opd} - c_i^{opvar} - c_i^{opland} - c_i^{opflb} \tag{4.5}$$

---

[4]Depreciation costs are estimated based on qualitative interviews.

whereas E_i = economic profit per hectare coffee; c_i^{opd} = interest for depulper; c_i^{opvar} = interest for variable costs; c_i^{opland} = opportunity costs for land[5] and c_i^{opflb} = opportunity costs for family labour. The interest rate used for evaluating the opportunity costs is 17%, based on a weighted average between the nominal interest rate for borrowed capital and the interest rate for savings[6]. We apply the estimation methods for opportunity costs of Reisch and Zeddies (1977) and Kay et al. (2004). The opportunity costs for family labour are valued with 60% of the salary farmers could obtain as casual day labourers. We choose this approach as hardly any off-farm employment exists and casual labour is not available all year round.

Additionally, we measure the break-even yield ($\bar{y}$) and break-even price ($\bar{p}$). The return to family labour per working day ($\bar{a}$) is obtained by dividing the accounting profit (eq. (4.4)) with the total amount of family labour person-days. A person-day is assumed to be eight hours which has been found in our study to be equivalent to the average working time on coffee farms in the region. The ratio of on-farm family labour per hectare coffee ($\bar{r}$) is calculated by divding the number of self-employed household members in agriculture with the total coffee area. The statistical measurement of profitability is based on equations (4.1) - (4.5), $\bar{y}, \bar{p}, \bar{a}$ and $\bar{r}$. Profitability is measured for all sampled households and weighted according to the proportion of sample size to total cooperative size in each producer group.

4.2.3. Measuring absolute and relative poverty

We apply the income method to measure absolute poverty. In order to measure relative poverty, we construct a poverty index. This section describes both methods.

Measuring absolute poverty

Absolute poverty measurements have traditionally relied on either measuring total income or total consumption expenditures of a household. As we are interested in coffee income and its contribution to the total income, we chose to collect income data of a household. Apart from the coffee data, additional income variables like off-farm income, remittances, pensions, crops sold and consumed, sold animals or animal products, and sales of firewood are used. Our approach has certain limitations. The value of sold animals is likely to be overestimated as production costs were not collected. The production costs of food crops are in part estimated based on secondary data from Aburto Sanchez (2008). The value of home consumption of animals and their products, an in-kind source of income,

---

[5]As some interviewees had difficulties estimating their land value, we use the mean value across all households.

[6]The average yearly interest rate for borrowed capital is 24%, for savings 3%. As most of the expenses in coffee production are covered by credit, we weight the rate for borrowed capital at 70% and for savings at 30%.

is omitted. However, this omission is likely to cause only a small error as there is – apart from chicken eggs – a very low amount of home-consumption of animal products observed. The equation used for total per capita income is

$$I_i^{pc} = \frac{1}{f_i} \sum_{j=1}^{7} z_{ij} + \sum_{k=1}^{7} (r_{ik} - c_{ik}) \qquad (4.6)$$

where I_i^{pc} = total per capita income; i = household index ($i = 1, \ldots, N$); f_i = family size; j = non-crop income type; z_{ij} = income of household i generated by income type j; k = crop type; r_{ik} = return of crop k; and c_{ik} = production costs for crop k.

According to the total per capita income (eq. (4.6)), it is possible to classify a coffee producing household above or below national and international poverty lines. We apply four absolute poverty lines. The national poverty line and the extreme poverty line for 2007 are based on the poverty lines from 2005 published in the 'Nicaragua Poverty Assessment' of the World Bank (2008). We adjust them with the Consumer Price Index (CPI) for 2007. The extreme poverty line is a food poverty line while the general poverty line includes expenditures such as housing, clothing and schooling. For 2007, the national poverty line is 442.6US$ and the extreme poverty line 246.8US$. We include the '$1-a-day' and '$2-a-day' poverty lines because these are the common poverty lines for international country comparisons (Van de Ruit and May, 2003). After being adjusted with the Purchasing Power Parity (PPP) for Nicaragua in 2007, the '$1-a-day' poverty line is equivalent to 200.6US$ per year; the '$2-a-day' poverty line to 320.95US$ per year. All poverty lines use income/expenditure as the only dimension to describe well-being.

4.2.4. Measuring relative poverty

Poverty's multidimensionality means that it cannot be exclusively measured by income or expenditures (Anand and Sen, 1997). Therefore, we also measure poverty using an index derived from indicators which cover different dimensions of poverty like quality of housing, food consumption, household assets and demographic data. Several studies have found that these proxy variables are reliable indicators of well-being (Van de Ruit and May, 2003; Zeller et al., 2006). These variables are aggregated, based on defined weights, into a univariate poverty index. The weights for poverty indices can be derived by several methods. One method is 'Principal Component Analysis (PCA)', which several authors have applied to derive so-called wealth or poverty indices with good results (Filmer and Pritchett, 2001; Van de Ruit and May, 2003; Zeller et al., 2006, 2003). We follow the method of measuring relative poverty described by Zeller et al. (2006) which uses PCA to determine the weights for a defined set of variables and to obtain a poverty score of each household. The model estimating the poverty index is calibrated for the control group of conventional coffee producers. Variables which show no impact are excluded. The final model is then applied to the whole sample. In the next step, the conventional control group is divided into terciles yielding three groups with poorest, less poor and least poor households. According to their poverty score, certified coffee households are classified into one of the three poverty groups. The quantitative data was triangulated with qualitative data from the focus group and individual interviews.

In order to identify how poverty developed among the three groups from 1997 to 2007, we used recall questions for these time periods as panel data does not exist. The recall questions were based on quantitatively measurable variables which are easily remembered like housing conditions, and the ownership of transport assets, major household appliances and (electronic) assets, animals and farm tools. While the data obtained by such long-term recall questions is likely to be less precise than the information regarding current asset status, the asset indicators chosen (e.g. TV, radio, number of rooms) were vividly remembered by all interviewees given their low asset levels.

To measure the changes of relative poverty over time, we include only those households that already existed in 1997, which decreases the sample size. To compare the relative poverty status over time, poverty indices are calculated for 1997, 2002 and 2007, allowing for flexible weights and flexible variables as the indicators for best describing relative poverty are likely to change over such a time period. This approach may be affected by a selection bias as those households which had left coffee production or the cooperatives since 1997 could not be included. It is difficult to control for this missing data and it could influence the result in either direction.

4.3. Results and discussion

The household characteristics of conventional, organic and organic-fairtrade certified producers do not show strong differences (Table 4.1). In general, small-scale coffee producing households are mainly headed by men in their forties, the majority of whom are able to read and write, but have not finished primary school. They take care of more than five household members, of which up to 50% are children. Organic-fairtrade farmers have the highest child dependency ratio. More than 20% of the household adults are illiterate and around 40% did not finish their primary school. Around 30% participated in secondary education, often the young adults, but only few finished high school or have tertiary education. Close to 50% of the household adults are self-employed in agriculture. Food crops are grown beside coffee and are usually consumed at home; surpluses are sold at local markets. Mainly male adults work in agriculture while women are more involved in domestic work. Apart from helping out in coffee production - especially at harvest time - some women also own tiny convenience shops. Off-farm work is more common among certified producers but is not a major income source. Conventional and organic certified producers gain around 60% of their available income from coffee production, organic-fairtrade certified producers slightly less. As total per capita income is not higher than 500US$

per year, the asset values are pretty low. Conventional farmers own more land and invest more in animals while organic-fairtrade farmers invest in transport assets. The standard deviation of some variables is very high as each group includes some very poor and some better-off farmers. This is the main reason for apparently different mean values being not statistically different. In the research region, smallholder coffee producers generally

Table 4.1.: Profile of conventional and certified smallholder coffee producing households, mean values for 2007

	Conventional [n=68]	Organic [n=108]	Organic-Fairtrade [n=105]
Share of female-headed household (%)‡	8.4 ± 28.0	17.6 ± 38.3	13.4 ± 34.2
Age of household head†	46.3 ± 13.0	43.5 ± 12.4	43.5 ± 11.3
Household head can read and write (%)‡	80.2 ± 40.2	76.1 ± 42.8	72.4 ± 44.9
Number of household members†	5.4 ± 1.8	5.7 ± 2.3	5.7 ± 2.1
Child dependency ratio†	0.8 ± 0.6	0.7 ± 0.6^{a}	0.9 ± 0.7^{a}
Share of illiterate adults (%)*‡	20.0 ± 25.8	20.9 ± 31.2	29.2 ± 34.9
Share of adults with < 6 years education (%)*†	38.8 ± 30.1	46.8 ± 30.8	42.6 ± 33.0
Share of adults with ≥ 6 and < 12 years education (%)*†	31.6 ± 22.0	33.2 ± 21.0	26.5 ± 19.0
Share of adults with ≥ 12 years education (%)*‡	11.0 ± 14.0^{e}	4.7 ± 12.5^{e}	9.4 ± 16.4
Share of adults self-employed in agriculture (%)‡	47.0 ± 17.9	48.7 ± 19.7	42.6 ± 22.6
Share of adults with off-farm work (%)‡	3.3 ± 11.8^{c}	6.9 ± 15.5	9.7 ± 20.3^{c}
Number of self-employed in agriculture‡	1.5 ± 0.8	1.8 ± 1.2^{a}	1.3 ± 0.8^{a}
Total land area (ha)‡	8.7 ± 10.2^{a}	5.4 ± 7.5^{a}	6.0 ± 6.6
Per capita value of animal assets (US$)‡	186.7 ± 754.0	103.9 ± 346.6	56.6 ± 133.1
Per capita value of transport assets (US$)‡	99.1 ± 266.8	69.9 ± 322.2	158.6 ± 828.4
Per capita value of major household assets (US$)‡	41.3 ± 83.4^{c}	18.8 ± 38.7^{c}	30.1 ± 57.7
Per capita value of yearly off-farm income (US$)‡	54.5 ± 145.9^{c}	43.9 ± 142.9^{a}	126.9 ± 543.5^{ca}
Per capita value of home consumption (US$)‡	68.1 ± 79.7	58.4 ± 52.4	48.9 ± 53.1
Total per capita income (US$)‡	498.5 ± 432.7^{e}	406.7 ± 390.3^{e}	499.6 ± 957.6
Share coffee income to total per capita income (%)‡	59.9 ± 28.6	63.1 ± 26.3^{e}	53.7 ± 48.1^{e}

Note: Superscript letters indicate a significant difference between two groups marked by the same letter. ^{a} or ^{b} indicate a significant difference at $p < 0.01$, ^{c} or ^{d} indicate a significant difference at $p < 0.05$, and ^{e} indicates significance at $p < 0.1$.

* The educational shares do not sum up to 100% as some adults went to primary school but have not learned to read or write.

† Normally distributed variables. ANOVA was used to test for statistical differences, followed by pairwise comparisons based on the Bonferroni post hoc test.

‡ Non-normally distributed variables. Kruskal-Wallis tests were used to test for statistical differences, followed by pairwise comparisons based on the Mann-Whitney post hoc test adjusted by the Bonferroni correction factor (Field 2005).

pursue low input production systems. Their coffee plantations are of mixed age with 20 year old trees next to rejuvenated or young coffee trees. The average green coffee yield in Nicaragua is 761.45kg/ha (IICA, 2004). Conventional and organic-fairtrade certified coffee yields are more than 50% below national average, yields from organic coffee on average 35% lower (Table 4.2). The qualitative interviews indicated that the main reasons for these differences are poorly managed plantations, insufficient fertilization and low planting densities. The higher yield levels of organic farmers in comparison to the other producers are unexpected. The differences cannot be contributed to the organic production system as such because the fairtrade certified producers have the same production system. Possible reasons can be differences in the applied quantities of organic fertilizers, better coffee management practices of organic farmers as well as a higher density of coffee trees per hectare. They relate to the ratio of land size and number of self-employed adults. Organic producers have the smallest mean coffee area but highest ratio of on-farm family labour per coffee hectare (see Table 4.3). They can manage their coffee plantations without having to invest much in hired labour and thus in labour supervision costs. This management could be explained by diseconomies of scale frequently found in family farms (Eastwood et al., 2010). Another explanation would be that smaller plots tend to receive more attention than larger coffee plots which can be described by the inverse relationship between farm size and land productivity (and thus the labour/land ratio) (Carter, 1984; Eastwood et al., 2010).

The certified cooperatives sell over 85% of their coffee in certified market channels. The average farm-gate coffee price differed significantly between conventional and organic as well as between conventional and organic-fairtrade farmers (Table 4.2). In comparison to conventional coffee prices, organic producers received 8% and organic-fairtrade producers 11% higher prices. The absolute difference is around 0.2US\$/kg[7] coffee which confirms our hypothesis (2a) of higher farm-gate prices paid in certified than in conventional market channels.

Due to the different yield levels, the revenue is significantly higher for the organic farmers while organic-fairtrade certified producers have only slightly higher revenues than their conventional colleagues. The certified coffee farmers know this problem as one organic-fairtrade certified producer explained *"[The price] is really good but more in our case it is not the price it is the yield we have per manzana[8], we don't have a good yield"* (organic-fairtrade producer in a focus group interview, 7 April 2008). We assumed that organic coffee production involves lower expenditures for purchased inputs (hypothesis (1)). This cannot be confirmed because total input costs do not differ significantly between the groups, although the mean input expenditures for organic and organic-fairtrade producers were lower than for conventional producers (Table 4.2). Many of the organic production processes are more laborious as shown by the significant differences in total person-days per ha in comparison to conventional production. The higher labour intensity was also frequently mentioned by farmers in the qualitative interviews. One organic-fairtrade pro-

---

[7]The exchange rate for Nicaraguan Córdoba against US\$ was 1US\$ = 18.96C\$ as of 31 August 2007.

[8]1 manzana = 0.705 hectare

Table 4.2.: Gross margins and profit calculations per ha of conventional and certified coffee production, 2007

	Conventional [n=68]	Organic [n=108]	Organic-Fairtrade [n=104]
Coffee area (ha)[‡]	3.2 ± 2.3^{ab}	2.2 ± 2.0^{a}	2.4 ± 2.6^{b}
Yield of green coffee (kg/ha)[‡]	365.9 ± 192.4^{e}	434.4 ± 253.8^{ea}	353.9 ± 176.7^{a}
Average farm-gate coffee price P_i^{aver} (US$/kg)[‡]	2.1 ± 0.2^{ab}	2.3 ± 0.2^{a}	2.3 ± 0.3^{b}
Revenue from coffee (US$/ha)[‡]	$\mathbf{762.7 \pm 395.3}^{a}$	$\mathbf{987.7 \pm 587.9}^{ac}$	$\mathbf{825.1 \pm 421.4}^{c}$
Chemical disease, pest, weed treatment[‡]	5.1 ± 13.0	0.0 ± 0.0	0.0 ± 0.0
Organic disease & pest treatment[‡]	1.5 ± 6.0^{ab}	4.1 ± 6.9^{a}	8.0 ± 22.2^{b}
Chemical fertilizer[‡]	32.3 ± 73.0	0.0 ± 0.0	0.0 ± 0.0
Organic fertilizer[‡]	0.1 ± 0.7^{a}	1.5 ± 9.4^{b}	7.5 ± 40.7^{ab}
Nursery and transport[‡]	8.6 ± 11.5	10.9 ± 18.0	8.6 ± 9.9
Total input costs (US$/ha)[‡]	47.6 ± 85.0	16.5 ± 22.6	24.1 ± 63.5
Harvest costs (incl. hired harvest workers)[‡]	129.7 ± 85.1	136.9 ± 97.7	141.8 ± 91.2
Hired labour (without labour employed in harvest) [‡]	81.5 ± 86.7	91.8 ± 152.8^{c}	141.7 ± 159.9^{c}
Total variable production costs (US$/ha)[‡]	258.7 ± 159.4	245.2 ± 201.8	307.6 ± 242.5
Gross margin from coffee (US$/ha)[‡]	$\mathbf{504.0 \pm 349.2}^{a}$	$\mathbf{742.5 \pm 514.2}^{ab}$	$\mathbf{517.5 \pm 328.2}^{b}$
Depreciation of depulper (US$/ha)[‡]	14.1 ± 10.1^{ab}	26.3 ± 27.0^{a}	20.8 ± 14.5^{b}
Accounting profit (US$/ha)[‡]	$\mathbf{489.9 \pm 346.0}^{a}$	$\mathbf{716.1 \pm 502.0}^{ab}$	$\mathbf{496.7 \pm 330.4}^{b}$
Interest of depulper[‡]	22.1 ± 16.0^{ab}	41.4 ± 42.5^{a}	32.7 ± 22.8^{b}
Interest for variable production costs for 6 months[‡]	11.0 ± 6.8	10.4 ± 8.6	13.1 ± 10.3
Land charge*[‡]	58.0 ± 0.0	58.0 ± 0.0	58.0 ± 0.0
Opportunity costs of family labour[‡]	207.6 ± 130.0^{c}	278.2 ± 183.4^{c}	246.1 ± 166.1
Economic profit (US$/ha)[‡]	$\mathbf{191.2 \pm 292.6}^{c}$	$\mathbf{328.2 \pm 394.5}^{ca}$	$\mathbf{146.8 \pm 298.1}^{a}$

Note: Superscript letters indicate a significant difference between two groups marked by the same letter. [a] or [b] indicate a significant difference at $p < 0.01$, [c] or [d] indicate a significant difference at $p < 0.05$, and [e] indicates significance at $p < 0.1$.

[‡] Non-normally distributed variables. The same significance tests as noted in Table 4.1 were applied. * Local interest rate for savings of 3% used as most producers obtained land for free from land distribution and renting of coffee areas is uncommon.

Table 4.3.: Break-even analysis, labour requirements and net coffee incomes of conventional and certified coffee production, 2007

	Conventional [n=68]	Organic [n=108]	Organic-Fairtrade [n=104]
Break-even yield at average farm-gate price (kg/ha)[‡]	124.6 ± 76.5	107.6 ± 88.2	133.5 ± 102.5
Break-even price at given hectare yield level (US$)[‡]	0.8 ± 0.5^e	0.7 ± 0.7^{ea}	0.9 ± 0.6^a
Return per family labour person-day (US$/day)[‡]	6.2 ± 6.4	5.4 ± 3.3	5.1 ± 3.4
No. of person-days used per ha[‡]	104.3 ± 52.2^{ac}	156.1 ± 84.3^{ad}	136.9 ± 65.3^{cd}
No. of hired person-days used per ha[*‡]	17.9 ± 22.3^c	23.0 ± 34.7^e	35.2 ± 40.1^{ce}
No. of family person-days used per ha[‡]	86.4 ± 49.7^a	133.1 ± 78.7^{ab}	101.7 ± 51.0^b
Ratio of on-farm family labour per ha coffee [‡]	0.7 ± 0.7^a	1.7 ± 2.4^{ab}	0.9 ± 0.8^b
Net income for whole coffee area (US$)[‡]	1416.9 ± 1373.7	1202.3 ± 1127.2	1361.2 ± 2227.4
Per capita net coffee income (whole area) (US$)[‡]	289.3 ± 302.3	240.2 ± 248.7	279.1 ± 462.2

Note: Superscript letters indicate a significant difference between two groups marked by the same letter. ^{a} or ^{b} indicate a significant difference at $p < 0.01$, ^{c} or ^{d} indicate a significant difference at $p < 0.05$, and ^{e} indicates significance at $p < 0.1$.

[‡] Non-normally distributed variables. The same significance tests as noted in Table 4.1 were applied.

[*] Labour for weeding and harvest not included when paid as lump sum.

ducer explained that *"one the one hand, the organic [coffee] is [cheaper] because one spends less but as it was said there is more work, one has to work more because one has to make these compost heaps"* (focus group interview, 16 April 2008). Organic farmers cover the additional labour requirements mainly by family labour, although they also hire more labour than conventional farmers. Organic-fairtrade certified farmers tend to use less family labour and hire more labour which can be explained by the shortage of family labour, their higher child dependency and location effects with higher opportunity costs (Table 4.1). The sum of total variable production costs shows no significant differences between the three producer groups. This was not anticipated. The additional labour requirements of organic production were assumed to be fully covered by available family labour which is only partly true. Certified coffee producing families, especially organic-fairtrade certified ones, face labour constraints and thus need to hire additional labour. Therefore, the hypothesis (1) needs to be rejected as neither input costs nor production costs are significantly lower of certified producers and labour costs are likely to rise. The comments of organic coffee producers confirm this: *"The [production] costs are much higher because you have to give better maintenance to the farm"* and *"if I think about that [the production costs] it is not profitable"* (two organic producers in a focus group interview, 24 April 2008).

In comparing the gross margins and accounting profit per hectare for the three producer groups, no difference between conventional and organic-fairtrade producers is found although organic producers have a significantly better gross margin and accounting profit than the other two groups. As the same production technique is used, we expected the accounting profits of organic-fairtrade producers to be at least similar if not better than organic producers. The contrasting finding between these two groups can be explained by differences in yield levels, coffee area, and the ratio of on-farm family labour per coffee hectare which determines also differences in costs for hired labour. The large standard deviations regarding gross margins and profits per hectare are commonly observed (Kilian et al., 2006). These deviations indicate that different farm management techniques are applied and that there is potential for further income gains across all producer groups. The break-even yield and break-even price for all groups is much below the obtained yield and price level, therefore production costs are covered even when small changes in prices or yields occur. The return to family labour is higher than the paid average salary of 3.77US\$/day, including in-kind payments obtained by casual day labourers. The investment of family labour in coffee production pays off among all three producer groups.

Positive economic profits per hectare coffee are derived on average by all producer groups, i.e. their coffee production is efficient. The highest efficiency is obtained by organic producers, the lowest by organic-fairtrade producers. The contrasting efficiency between organic and organic-fairtrade producers cannot be related to the organic production system. The standard deviations point out that there are also inefficient producers. This apparent non-rationality of continuing coffee production could be explained rationally by further considering risk aversion, loss of social status and high costs for switching to other land uses.

Whether a household can generate an income above the poverty line with coffee production depends on the accounting profit for the whole coffee area, the net income, in relation to the family size. The net income for the whole coffee area is highest for conventional farmers, followed by organic-fairtrade certified farmers and lowest for organic farmers (Table 4.3). The per capita net coffee income follows the same order. Net coffee income for the whole area and per capita is not significantly different between the three producer groups. The per capita net coffee income in all producer groups is not high enough to enable farm households to meet all basic needs since per capita coffee incomes are below the national and the international '\$2-a-day' poverty line. As an organic-fairtrade producer puts it: *"We have many years working organic [...] we know that until now we have not, how to say it, better economic resources"* (focus group interview, 7 April 2008). Therefore, we anticipate the rejection of our hypothesis (2b) regarding per capita net coffee income. That certified farmers do not have higher net coffee incomes for the whole coffee area and per capita can be explained by their higher labour requirements due to organic production, and thus costs, which offset saved input costs. This is especially relevant for organic-fairtrade producers compared to conventional farmers. That organic producers fare slightly worse than conventional farmers can be additionally explained by their smaller coffee area.

Due to non-coffee income sources, total household income needs to be additionally considered when identifying the poverty status of a household. On average, only organic producers have a per capita income below the national poverty line (see Table 4.1). The other two groups have an average per capita income above the different poverty lines. As mean values do not reflect the heterogeneity within producer groups, producers are categorized according to their per capita incomes above or below the various poverty lines (Table 4.4). Compared to one-third of conventional producers, 45% of the organic and organic-fairtrade certified producers have per capita incomes below the extreme poverty line – which means that they cannot cover their food requirements. An organic producer explains that by the time the cooperative has sold all the coffee and paid their producers *"now it is sure that the coffee comes at a good price but now everything is expensive. Thus we, the poor, always change for the worse, you understand? Of course, for us poor everything is failing although we try. Sometimes we say [...] that it is better to sell the coffee at harvest time, although we will give it away for nothing, but we will buy cheaper beans and maize"* (organic producer in an interview, 18 May 2008). Around two-thirds of all producers are classified as poor according to the national poverty line. More organic and organic-fairtrade certified producer households than conventional producers have incomes below the poverty lines but no significant difference exists between the three groups. As the absolute poverty levels were only based on the income dimension, we

Table 4.4.: Percent of conventional and certified coffee producer households below the poverty line in 2007

	Conventional (%)	Organic (%)	Organic-fairtrade (%)
National extreme poverty line	30.9	44.4	44.8
National poverty line	60.9	71.3	68.6
World Bank '$1-a-day'[e]	23.5	38.9	36.2
World Bank '$2-a-day'[e]	39.7	56.5	56.2

Note: [e] indicates significance between the three groups at $p < 0.1$ using χ^2 tests.

further analysed the relative poverty levels based on different dimensions of poverty. The variables and their component loadings used for constructing the poverty index of 2007 are shown in Table 4.5. The variables cover several dimensions of poverty, such as quality of housing, food consumption, household assets and demographic data. Only one food variable could predict relative poverty differences between the groups. This was expected as chronic hunger and severe undernutrition is not so common among coffee farmers although food insecurity still persists. After calibration, the goodness of fit of the PCA model for all data is indicated by a Kaiser-Meyer-Olkin measure of sampling adequacy of 0.741, a highly significant Bartlett's Test of sphericity ($p < 0.000$), an eigenvalue of the first component of 4.167 and an explained variance of the model of 26%. The mean poverty index for conventional producers was 0.25571 ± 1.36924, for organic producers -0.10180 ± 0.82492 and for organic-fairtrade producers -0.01082 ± 0.84238. Post-hoc tests show that organic farmers are relative poorer than conventional farmers at $p < 0.05$. There is no significant difference between the other producer groups. The classification

Table 4.5.: Variables and component loadings of the relative poverty index, 2007

	Component loading[a]
Family structure	
Percent of literate household adults	0.299
Share of adults with complete secondary school	0.362
Share of adults self-employed in agriculture	−0.356
Share of adults self-employed in off-farm enterprise	0.479
Food consumption	
Times cheese served in last 7 days	0.485
Housing	
Floor material now	0.365
Cooking material now	0.516
Light source now	0.525
No of rooms per person	0.402
Assets	
Value of vehicles	0.418
Value of TVs	0.782
Value of DVD/VHS	0.579
Value of Radio	0.631
Value of mobile phones	0.615
Value of gas/electric oven	0.644
Value of storage room	0.452

Note: [a]Component loading ranges from -1 to $+1$

of conventional, organic and organic-fairtrade certified households in the three poverty groups, poorest, less poor and least poor, shows that more organic and organic-fairtrade certified households are in the poorest group than conventional households. To compare against the terciles of conventional households in each group, nearly half of the organic

(49%) and organic-fairtrade (46%) households are in the poorest group. Differences in the less poor group are minor. Only 18% of organic producers are categorized as least poor, compared to 27% of organic-fairtrade producers and 33% of conventional producers. By and large, the comparison of relative poverty among the three groups confirms the above results with respect to absolute poverty. That coffee certifications as such do not help smallholders to earn incomes above the poverty line is confirmed by the qualitative interviews with key-persons. As a NGO-expert working in the smallholder coffee sector states: *"We need to see what [coffee producers and their cooperatives] can do in addition to coffee, because you have producers which will not even escape the situation of poverty in which they are now even if the coffee prices reaches 200US$ el quintal*[9]*. If they do not have a diversification strategy and see how they can create other alternatives to generate income opportunities, so much to fair trade [...], the people will not escape their situation - producers with three manzanas which each produces four to five quintales [of coffee] and with seven children - not even if they earn 250US$ [per quintal] or more. In addition you have the problem that they don't have access to education, you have to pay for that [...]. All this sums up and in the end you do not have the possibility to escape the poverty level in which you are"* (key-person interview, 13 May 2008).

The question remains how poverty levels have developed in the past ten years. General living standards in Nicaragua have improved in the last ten years, as demonstrated by the increase in Human Development Index values from 0.597 in 1995 to 0.699 in 2007 (UNDP, 2009). Our data also show that all producer groups have developed positively and improved their housing conditions as well as asset composition from 1997 to 2007. To identify the trend in relative poverty among the three producer groups, poverty indices were calculated for the years 1997, 2002 and 2007.

We hypothesized that ten years ago conventional and certified farmers had equal poverty levels; the civil war only ended in 1990 and the country was still recovering from the vast destructions and losses (Ruben and Zerman, 2005). In 1997, there were less certified than conventional producers in the poorest group, more in the less poor and similar numbers in the least poor group (Table 4.6). The absence of statistical differences suggests that groups were rather homogeneous. Since 1997, the situation of organic farmers deteriorated and they became relatively poorer than the other groups in 2002 and 2007. The relative poverty levels of organic-fairtrade producers in 2002 reveal that, when conventional prices dropped below production costs during the coffee crisis, they were relatively less poor than the other two producer groups – an effect which may be contributed to the price stabilization through the guaranteed minimum price of fairtrade, described in other studies (Bacon, 2005; Raynolds et al., 2004). However, this positive trend reversed, and in 2007, organic-fairtrade certified farmers are predominately found in the poorest poverty group. They are relatively poorer than conventional farmers but not at statistically significant levels (Table 4.6). The hypothesis (3) that relative poverty levels of

---

[9]1 quintal = 46kg. The mentioned coffee price converts to 4.35US$/kg.

44

households participating in organic and organic-fairtrade developed more positively than the conventional producers has to be rejected. While not being free of potential selection bias, we see the result as a challenge to conventional assumptions and as an interesting starting point for further research on long-term poverty effects of certification. Given the

Table 4.6.: Comparison of relative poverty levels in 1997, 2002 and 2007; percent of conventional and certified households in three poverty groups

| | Household grouping | | | | | | | | |
| | 1997 | | | 2002 | | | 2007 | | |
	Poorest	Less poor	Least poor	Poorest	Less poor	Least poor	Poorest	Less poor	Least poor
Conventional [n=61]	32.8	34.4	32.8	34.4	32.8	32.8	34.4	32.8	32.8
Organic [n=84]	26.2	41.7	32.1	41.7	34.5	23.8	51.2	32.1	16.7
Organic-fairtrade [n=92]	25.0	42.4	32.6	33.7	28.3	38.0	42.4	29.3	28.3

Note: To test for significant difference between the three groups χ^2 tests were used.

severe impact of the coffee crisis in Nicaragua, it is surprising that the relative well-being of organic-fairtrade producers during the crisis results in no lasting benefit for organic-fairtrade producers as currently they are relatively poorer than conventional producers. Conventional producers, who even had a slightly higher relative poverty level in 1997, suffered more in the coffee crisis and yet are slightly less relative and absolute poor than the organic and organic-fairtrade producers. More research is needed to understand why organic-fairtrade producers managed the coffee crisis better but do not develop equally well as conventional producers during times of good coffee prices.

4.4. Conclusion

In this paper, we first analysed if farm-gate coffee prices are higher in certified chains and lead to increasing net coffee income compared to conventional production. Second, we identify whether certified coffee producers are more or less poor than conventional producers with similar socio-economic characteristics and whether poverty levels have changed over time.

Organic-fairtrade coffee is found to achieve the highest farm-gate prices, followed by organic coffee in comparison to conventional prices. Organic production processes require fewer purchased inputs but are more laborious. Due to constrained availability of family labour, additional labour has to be hired which offsets saved input costs. The higher prices of certified coffees compensate for production costs but fail to increase per hectare gross margins and profits in the case of organic-fairtrade farmers compared to conventional produces. Organic producers have higher yields and thus experience an increase in per hectare gross margins and profits. Due to smaller coffee areas and large family size, the increase in gross margins does not result in improved per capita net coffee incomes for organic certified producers. Also organic-fairtrade certified producers do not have higher per capita net coffee incomes than conventional producers.

We conclude that the profitability of the organic certified production system is not clear cut; there is a trend that organic but not organic-fairtrade certified producers have higher gross margins and profits than conventional farmers in northern Nicaragua. Our study shows that higher farm-gate prices do not lead necessarily to higher per capita net coffee income, as yield levels, production costs, family and land size, as well as labour availability play important roles. Further research comparing several coffee producing countries could more specifically identify factors and conditions which determine economic success of certification schemes.

In comparison with previous literature that mainly approached poverty through qualitative studies, we measure poverty based on quantitative data. Among certified producers, a higher share of households is grouped below the poverty line than among conventional producers. This may indicate that the organic and organic-fairtrade coffee certification as such does not help northern Nicaraguan coffee farmers to earn a coffee income above the poverty line or to make them better off than their conventional fellow men. Moreover, we find that over a period of ten years, organic certified producers became relatively poorer. Organic-fairtrade certified producers first improved their relative poverty status and were relatively better off than conventional producers during the coffee crisis. After the crisis, the relative poverty levels of organic-fairtrade producers deteriorated compared to conventional producers. Yet, we cannot rule out the possibility that the results are influenced by selection bias. This paper does not provide for a causal econometric impact analysis and such studies are strongly suggested for further research to test the claims that certified coffee production contributes to poverty reduction. Quantitative socio-economic evaluations of certification schemes should ideally be done in different institutional settings with producer groups of various countries and should include panel data. With such a design, the institutional, policy and market settings could also be compared in order to identify the conditions in which certifications are likely to work or not. Although our sample design does not allow for generalizations beyond the study area of northern Nicaragua the results are likely to be representative for coffee growers in higher altitudes in Nicaragua as production conditions are similar in these areas.

We recommend that the policy focus of government and donors should move from certification schemes to investments in the farm and business management skills of producers as well as the establishment of public extension and production support systems. There is a need for an efficient public extension system equipped with adequate financing as cooperatives do not have sufficient funds of their own to deliver these services; so far cooperatives depend only on short-term external funding. Furthermore, we propose public support for cooperatives, for example state credit at market conditions, to remove cooperatives' liquidity constraints at harvest time or to enable them to improve their credit services to farmers. Individual land titling of former collective land belonging to the cooperatives should be eased and implemented more quickly. Cooperatives should be supported through policies and funds during this process as they face a high administrative burden. Organic certified cooperatives could also think about creating their own central organic fertilizer production to be sold to their members as they often face constraints in producing sufficient fertilizer for their fields. We conclude that coffee yields, profitability and efficiency need to be increased, as prices for certified coffee cannot compensate for low productivity, land or labour constraints.

Acknowledgements   Thanks are due to Thomas Oberthür, Lutz Göhring and Nicole Schönleber for their invaluable discussions and comments. The funding for field research by the 'Advisory Service on Agricultural Research for Development (BEAF)' of the 'Deutsche Gesellschaft für Technische Zusammenarbeit (GTZ)' and the logistic support by the 'International Center for Tropical Agriculture (CIAT)' in Colombia and Nicaragua, especially by Axel Schmidt and Evelyn López, is gratefully acknowledged. We are grateful to the three anonymous reviewers for their useful comments.

4.5. Bibliography

Aburto Sanchez, E., 2008. From subsistence to market oriented livestock smallholder development in Nicaragua and Honduras. Vol. 94 of Farming and Rural Systems Economics. Margraf, Weikersheim.

Anand, S., Sen, A., 1997. Concepts of human development and poverty: A multidimensional perspective. Tech. rep., United Nations Development Programme.

Arnould, E. J., Plastina, A., Ball, D., 2009. Does fair trade deliver on its core value proposition? Effects on income, educational attainment, and health in three countries. Journal of Public Policy & Marketing 28 (2), 186–201.

Bacon, C., 2005. Confronting the coffee crisis: Can fair trade, organic and specialty coffees reduce small-scale farmer vulnerability in northern Nicaragua? World Development 33 (3), 497–511.

Bacon, C., Ernesto Mendez, V., Gomez, M. E. F., Stuart, D., Flores, S. R. D., 2008. Are sustainable coffee certifications enough to secure farmer livelihoods? The Millennium

Development Goals and Nicaragua's fair trade cooperatives. Globalizations 5 (2), 259–274.

Basu, A. K., Chau, N. H., Grote, U., 2003. Eco-labeling and stages of development. Review of Development Economics 7 (2), 228–247.

Becchetti, L., Costantino, M., 2008. The effects of fair trade on affiliated producers: An impact analysis on Kenyan farmers. World Development 36 (5), 823–842.

Bolwig, S., Gibbon, P., Jones, S., 2009. The economics of smallholder organic contract farming in tropical Africa. World Development 37 (6), 1094–1104.

Bray, D. B., Sánchez, J. L. P., Murphy, E. C., 2002. Social dimensions of organic coffee production in Mexico: Lessons for eco-labeling initiatives. Society & Natural Resources 15 (5), 429–446.

Carter, M. R., 1984. Identification of the inverse relationship between farm size and productivity: An empirical analysis of peasant agricultural production. Oxford Economic Papers 36 (1), 131–145.

Daviron, B., Ponte, S., 2005. The coffee paradox. Global markets, commodity trade and the elusive promise of development. Zed Books, London and New York.

Eastwood, R., Lipton, M., Newell, A., 2010. Farm size. In: Pingali, P., Evenson, R. (Eds.), Handbook of Agricultural Economics. Vol. 4. Elsevier, Oxford & Amsterdam, pp. 3323–3397.

Filmer, D., Pritchett, L. H., 2001. Estimating wealth effects without expenditure data - or tears: An application to educational enrollments in states of India. Demography 38 (1), 115–132.

Fitter, R., Kaplinsky, R., 2001. Who gains from product rents as the coffee market becomes more differentiated? A value-chain analysis. IDS Bulletin 32 (3), 69–82.

Gobbi, J. A., 2000. Is biodiversity-friendly coffee financially viable? An analysis of five different coffee production systems in western El Salvador. Ecological Economics 33 (2), 267–281.

González, A. A., Nigh, R., 2005. Smallholder participation and certification of organic farm products in Mexico. Journal of Rural Studies 21 (4).

Gordon, C., Manson, R., Sundberg, J., Cruz-Angón, A., 2007. Biodiversity, profitability, and vegetation structure in a Mexican coffee agroecosystem. Agriculture, Ecosystems & Environment 118 (1-4), 256–266.

IFOAM, 2006a. Organic agriculture and food security. International Federation of Organic Agriculture Movements (IFOAM), Bonn, accessed 20 July 2006.
URL `http://www.ifoam.org/organic_facts/food/pdfs/Food_Security_Leaflet.pdf`.

IFOAM, 2006b. Organic agriculture and rural development. International Federation of Organic Agriculture Movements (IFOAM), Bonn, accessed 15 December 2010.
URL `http://www.ifoam.org/growing_organic/3_advocacy_lobbying/eng_leaflet_PDF/Rural_Development.pdf`.

IFOAM, 2009a. Definition of organic agriculture. International Federation of Organic Agriculture Movements (IFOAM), Bonn, accessed 29 March 2010.
URL `http://www.ifoam.org/growing_organic/definitions/doa/index.html`.

IFOAM, 2009b. The principles of organic agriculture. International Federation of Organic Agriculture Movements (IFOAM), Bonn, accessed 15 December 2010.
URL `http://www.ifoam.org/about_ifoam/principles/index.html`.

IICA, 2004. Cadena agroindustrial del café en Nicaragua. Instituto Interamericano de Cooperación para la Agricultura (IICA), Managua.

Kay, R. D., Edwards, W. M., Duffy, P. A., 2004. Farm management, 5th Edition. McGraw-Hill, New York.

Kilian, B., Jones, C., Pratt, L., Villalobos, A., 2006. Is sustainable agriculture a viable strategy to improve farm income in Central America? A case study on coffee. Journal of Business Research 59 (3), 322–330.

Lewin, B., Giovannucci, D., Varangis, P., 2004. Coffee markets: New paradigms in global supply and demand. Agriculture and Rural Development Discussion Paper No. 3. World Bank, Washington D.C.

Linton, A., 2008. A niche for sustainability? Fair labor and environmentally sound practices in the specialty coffee industry. Globalizations 5 (2), 231–245.

Murray, D. L., Raynolds, L. T., 2006. The future of fair trade coffee: Dilemmas facing Latin America's small-scale producers. Development in Practice 16 (2), 179–191.

Mutersbaugh, T., 2002. The number is the beast: A political economy of organic-coffee certification and producer unionism. Environment and Planning A 34, 1165–1184.

Mutersbaugh, T., 2005. Fighting standards with standards: Harmonization, rents, and social accountability in certified agrofood networks. Environment and Planning A 37 (11), 2033–2051.

Nigh, R., 1997. Organic agriculture and globalization: A Maya associative corporation in chiapas, mexico. Human Organization 56 (4), 427–436.

Philpott, S. M., Bichier, P., Rice, R., Greenberg, R., 2007. Field-testing ecological and economic benefits of coffee certification programs. Conservation Biology 21 (4), 975–985.

Raynolds, L. T., 2009. Mainstreaming fair trade coffee: From partnership to traceability. World Development 37 (6), 1083–1093.

Raynolds, L. T., Murray, D., Taylor, P. L., 2004. Fair trade coffee: Building producer capacity via global networks. Journal of International Development 16 (8), 1109–1121.

Reisch, E., Zeddies, J., 1977. Einführung in die landwirtschaftliche Betriebslehre. Bd. 2. Spezieller Teil. Ulmer, Stuttgart.

Rice, R. A., 2001. Noble goals and challenging terrain: Organic and fair trade coffee movements in the global marketplace. Journal of Agricultural and Environmental Ethics 14 (1), 39–66.

Ruben, R., Zerman, Z., 2005. Why Nicaraguan peasants stay in agricultural production cooperatives. Revista Europea de Estudios Latinoamericanos y del Caribe 78, 31–47.

Slob, B., 2006. A fair share for coffee producers. In: Osterhaus, A. (Ed.), Business unusual. Successes and challenges of fair trade. FLO, IFAT, NEWS, and EFTA, Newcastle-upon-Tyne, pp. 121–139.

Taylor, P. L., Murray, D. L., Raynolds, L. T., 2005. Keeping trade fair: Governance challenges in the fair trade coffee initiative. Sustainable Development 13 (3), 199–208.

Torre, A., 2002. The traditional economic approach to measuring economic profit. In: Standfield, K. (Ed.), Intangible Management. Academic Press, San Diego.

UNDP, 2009. Human development report 2009. United Nations Development Program (UNDP), New York.

Utting-Chamorro, K., 2005. Does fair trade make a difference? The case of small coffee producers in Nicaragua. Development in Practice 15 (3-4), 584–599.

Valkila, J., 2009. Fair trade organic coffee production in Nicaragua – sustainable development or a poverty trap? Ecological Economics 68 (12), 3018–3025.

Van de Ruit, C., May, J., 2003. Triangulating qualitative and quantitative approaches to the measurement of poverty: A case study in limpopo province, south africa. IDS Bulletin 34 (4).

Van der Vossen, H. A. M., 2005. A critical analysis of the agronomic and economic sustainability of organic coffee production. Experimental Agriculture 41 (4), 449–473.

Willer, H., Yussefi, M. (Eds.), 2007. The world of organic agriculture. Statistics and emerging trends 2007, 9th Edition. International Federation of Organic Agriculture Movements (IFOAM) and Forschungsinstitut fuer biologischen Landbau (FiBL), Bonn.

Wills, C., 2006. Fair trade: What's it all about? In: Osterhaus, A. (Ed.), Business unusual. Successes and challenges of fair trade. FLO, IFAT, NEWS, and EFTA, Newcastle-upon-Tyne, pp. 7–27.

Wollni, M., Zeller, M., 2007. Do farmers benefit from participating in specialty markets and cooperatives? The case of coffee marketing in Costa Rica. Agricultural Economics 37 (2-3), 243–248.

World Bank, 2008. Nicaragua poverty assessment. Volume 1: Main Report. No. 39736-NI. World Bank, Washington D.C.

Zeller, M., Sharma, M., Henry, C., Lapenu, C., 2006. An operational method for assessing the poverty outreach performance of development policies and projects: Results of case studies in Africa, Asia, and Latin america. World Development 34 (3), 446–464.

Zeller, M., Wollni, M., Abu Shaban, A., 2003. Evaluating the poverty outreach of development programs: Results from case studies in Indonesia and Mexico. Quarterly Journal of International Agriculture 42 (3), 371–383.

Chapter 5

Justified Hopes or Utopian Thinking? The Suitability of Coffee Certification Schemes to Increase Food Security for Small-Scale Producers

TINA BEUCHELT, MANFRED ZELLER AND THOMAS OBERTHÜR

This chapter has been submitted in February 2011 as article for publication to the Quarterly Journal of International Agriculture.

Abstract

Coffee certifications are promoted by governments and donors assuming that it contributes to poverty reduction and increases food security of poor small-scale farmers. There are few studies who verify these claims. This research analyzes the food security status and financial situation of conventional and certified farmers. Combining qualitative and quantitative research we compare conventional, organic, and Organic-Fairtrade certified small-scale coffee producers in Nicaragua.

Our results indicate that certification schemes have a low impact on food security or finances. Given low yield levels, certified coffee prices are insufficient to cover living and farm expenditures. Necessary financial needs are covered by credits. Many farmers are trapped in a vicious cycle of indebtedness. Coffee certifications alone do not help smallholder coffee growers to escape poverty in Nicaragua. Policies need to address entrepreneurial and financial skills of farmers as well as increase yield levels through e.g., research and extension.

Keywords:

Cooperatives; Fairtrade; food security; indebtedness; organic

JEL classification:

Q01 - Sustainable development

Q12 - Micro Analysis of Farm Firms, Farm Households, and Farm Input Markets

5.1. Introduction

Coffee is the main income source for 20-25 million families in East African, South Asian and Latin American hillsides (Gresser and Tickel, 2002), where most coffee is grown on farms smaller than five hectares (Fitter and Kaplinsky, 2001). In Nicaragua, the second poorest country in Latin America, coffee contributes 24% to total exports earnings; between 20%-40% of the rural labor force is employed in the coffee production (Lewin et al., 2004; Vakis et al., 2004).

While coffee is an important income source for developing countries and their small-scale producers, coffee prices are highly volatile and crises are common (Cashin et al., 2002). The last worldwide coffee crisis from 1998/99-2002/03 affected producers income the most (ICO, 2004) and between 2000 and 2001, prices, adjusted for inflation, dropped to their lowest level in 100 years (Varangis et al., 2003). In many regions these prices were below the production costs (Fitter and Kaplinsky, 2001; Raynolds et al., 2004). Coffee farms were neglected or abandoned. As living standards declined, social unrest and insecurity in hillside coffee producing regions grew (ICO, 2004; Varangis et al., 2003). Smallholders have been among the hardest hit by this price decline. Between 1998 and 2001, poverty rates of Nicaraguan smallholder coffee farmers increased by 2% while the poverty rate among rural households dropped by 6% (Lewin et al., 2004).

Paradoxically, at the same time the coffee market in importing countries flourished, with the value of the retail market doubling (ICO, 2004). In recent years organic or Fairtrade coffees became popular among roasters and consumers. This is driven by quality aspects but also by an increasing social, environmental or health consciousness (Daviron and Ponte, 2005; Rice, 2001). In 2006, the US reported growth rates of 56% for organic coffee imports and 33% for Fairtrade coffee (Giovannucci and Villalobos, 2007). Fair trade coffee consumption worldwide is growing annually 20% (FLO, 2007).

Differentiated coffees, such as certified or gourmet coffee become an interesting alternative for farmers, as markets tend to offer more stable and even higher prices for these coffees (Bacon, 2005; Daviron and Ponte, 2005; Lewin et al., 2004; Wollni and Zeller, 2007). Thus, governments, NGOs and international donors promote the marketing of coffee through group-based, certified market channels as a viable business model for poor small-scale farmers (Linton, 2008).

Cooperatives are the main producers of Fairtrade and organic certified coffee (Rice, 2001). The standards for organic coffee depend on the importing country and the certification label. The International Federation of Organic Agriculture Movements (IFOAM) defines several principles on which organic agriculture is based. Organic agriculture should enhance the health of soils, plants, animals and humans, the use of synthetic agro-chemical inputs is not allowed. It is a holistic approach which aims at a sustainable resource use and requires the interaction of humans to be fair at all levels and to all parties (IFOAM, 2009a,b). Since the organic certification is too costly for an individual small-scale producer, farmers form producer groups or join cooperatives to obtain group certification (Rice, 2001).

Fair trade is defined by the World Fair Trade Organization (WFTO) as *"a trading partnership, based on dialogue, transparency and respect, that seeks greater equity in international trade"* (WFTO, 2009). Also Fairtrade standards follow several key principles like the creation of opportunities for economically disadvantaged producers; payment of a fair price which for coffee means a guaranteed minimum price and a premium[1]; pre-financing and 'ethical' trade relations; transparency and accountability; capacity building; encouragement of better environmental practices and gender equity (WFTO, 2010). Fairtrade standards require that coffee producers are small, family-based growers organized into politically independent democratic associations (FLO, 2008).

Both Fairtrade and organic certification claim to contribute to poverty reduction and food security in developing countries. They boost rural development through enhancing governance, creating employment opportunities, maintaining a healthy environment and enhancing social capital (IFOAM, 2006; Osterhaus, 2006). While there is a growing body of literature regarding the effects of participation in certification schemes, many studies (Bacon, 2005; Murray and Raynolds, 2006; Raynolds et al., 2004; Wollni and Zeller, 2007) focus on farm-gate price differences but do not consider simultaneously the direct and indirect costs of participation. There are only very few studies that consider the socio-economic context of farmers and investigate the poverty alleviation and food security effects of certification schemes (Arnould et al., 2007; Bacon et al., 2008). We summarize the findings of these studies next.

Wollni and Zeller (2007) find for Costa Rica that participation in specialty coffee marketing channels as well as in cooperatives leads to higher farm-gate prices compared to conventional channels. During the coffee crisis, Fairtrade farmers in Mexico, Guatemala and El Salvador received two to three times higher prices than conventional farmers (Raynolds et al., 2004). Bray et al. (2002) show for Mexico that higher prices for organic coffee were offsetting higher production costs and farmers benefited from participation. In contrast, Mutersbaugh (2002) demonstrates that organic certification was only successful when farmers had already high yields levels. Additionally, organic price premiums have declined over the last 20 years "even as quality has increased, mainly because supply has grown" (Daviron and Ponte, 2005, p 173). Despite participating in differentiated markets farmers report a decline in their quality of life over the last few years (Bacon, 2005). Arnould et al. (2007) find only moderate positive impacts from participation in Fairtrade certification schemes in Nicaragua, Guatemala and Peru. Similar results are obtained by Bacon et al. (2008) who discover positive impacts regarding education and infrastructure investments but continuing low incomes, emigration and food insecurity. Summarizing, effects of certification schemes on smallholders are not clear-cut. It is crucial for producer organizations, policy makers and international donors to have a better scientific evidence base for their support of certification schemes to reduce food insecurity and poverty. This research seeks to complement the existing literature through focusing on food security and the financial situation of smallholder coffee producers. First, the research aims at

---

[1] From July 2008 onwards, the fair trade minimum price for washed Arabica conventional coffee is 2.75US$/kg (Free on Board - FOB), the organic differential is 0.44US$/kg and the social premium 0.22US$/kg. In Nicaragua, Arabica coffee is the common coffee variety.

analyzing the labor burden, financial, food and health situation throughout a year of conventional, organic and Organic-Fairtrade certified producers. Second, it identifies the food purchasing and consumption patterns of conventional, organic and Organic-Fairtrade certified producers. Third, it is analyzed which role financial and liquidity constraints play in coffee production and the livelihood of conventional, organic and Organic-Fairtrade certified smallholder coffee producers. We apply a combination of qualitative and quantitative research methods to small-scale producers that are organized in conventional, organic and Organic-Fairtrade certified cooperatives in northern Nicaragua.

5.2. Conceptual framework and research design

First, the conceptual framework is explained. This is followed by a description of the research area and sampling design.

5.2.1. Conceptual framework

Departing from the sustainable livelihoods approach of Scoones (1998), we have developed our own analytical framework (Figure 5.1). There are five asset categories or types of capital upon which the livelihoods of smallholder coffee producers are based: human, natural, financial, social and physical capital. These assets determine their access to cooperatives, certification and credit. In return, the markets, institutions and cooperatives influence the livelihood assets of small-scale coffee producers and form the basis of the producer's production and certification strategy. The decision of a farmer to participate in cooperatives and in certification schemes is assumed to depend on the utility a farmer attributes to participation. Where there are substantial benefits to be obtained through collective action, households will get involved (Varughese and Ostrom, 2001). The production and marketing performance affects the income and poverty status, food security levels and knowledge levels. It further influences the livelihood assets and vulnerability of farmers. Coffee price volatility has long been the most predominant threat to small-scale farmers and depends heavily on global coffee production.

5.2.2. Research area and research design

The research was conducted in northern Nicaragua in a region with similar coffee growing characteristics. The majority of coffee farms were between 900m and 1300m above sea level. We used a combination of qualitative and quantitative research. Seven cooperatives were randomly selected and grouped according to conventional, organic or Organic-Fairtrade certified cooperatives. The conventional cooperatives constituted the control group. Cooperatives had to be certified for a minimum of five years.

The quantitative household data was collected in 2007. The coffee producing members of the cooperatives were again randomly selected and 327 households were surveyed with

Figure 5.1.: Conceptual framework identifying factors that affect the performance of conventional, organic and Organic-Fairtrade certified producers

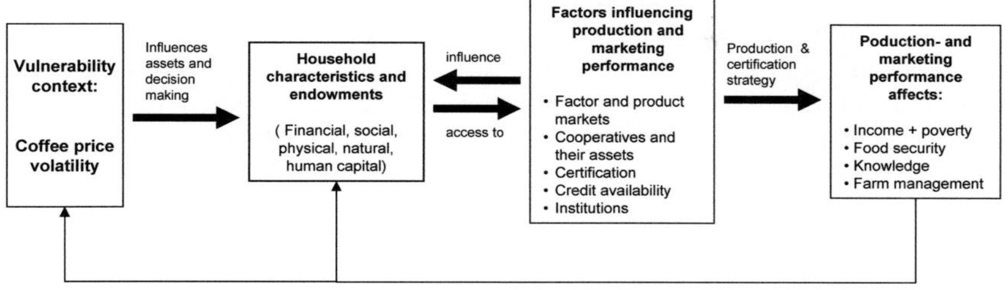

Source: Own illustration.

a structured questionnaire. Qualitative data was collected in 2007 and 2008. 48 semi-structured key-person interviews were conducted with leaders and staff of cooperatives, exporters, roasters, and researchers. This was complemented by 21 semi-structured and unstructured focus group discussions and 33 semi-structured interviews with small-scale coffee producers.

5.3. Results and discussion

"Coffee pays for everything, from the shoes to the top, [it pays] liquor for those who like liquor, women, for those who like women. Everything comes from the same coffee. That is why it never gives us enough to improve our living" (conventional producer, April 2008).[2]

5.3.1. Organic and Fairtrade coffee production

The citation indicates the important role coffee plays in the daily life of small-scale producers. Despite this importance, their conventional and Organic-Fairtrade certified coffee yields are more than 50% below the national average, organic coffee yields 40% lower. Highest yields (on average around 480kg/ha) are achieved by organic producers but yield levels vary, like for conventional and Organic-Fairtrade certified producers, between the cooperatives (ranging from 293kg/ha to 516kg/ha). The qualitative results indicated that main reasons for the low yields are badly managed plantations and low planting densities. In part, these are still consequences from the last coffee crisis during which producers had neglected their farms.

---

[2] All citations from producers are translated by the authors.

Organic farming requires a complex management and understanding of the ecological system. Farmers explained that a few group training sessions, as offered in one cooperative, are not enough for the, to learn organic farming methods properly. In certified cooperatives with less extension, farmers complained that they lack sufficient knowledge on organic production. Since most farmers lack money to purchase sufficient synthetic production inputs entry in organic certification schemes is eased. Yet, maintaining soil fertility is a major problem on organic farms. Coffee producers stated that their soils are quite exhausted, so yields depend a lot on fertilization. Organic fertilizer is very labor intense to be produced on the farm. Raw materials are scarce in the higher altitudes and, thus, often need to be purchased and transported over longer distances, making organic fertilizer relatively costly. Its purchase is also expensive. When the additional premium obtained for organic coffee needs to be in invested in employment of day-laborers or in input costs, organic farming becomes less attractive for the smallholder producers. Therefore, farmers often opt not to apply all necessary organic production techniques, for example they reduce fertilization. As one producer explained: *"We do not apply all organic practices which they tell us, we do not apply them, because financing is lacking"* (Organic-Fairtrade producer, April 2008).

The main reason why producers participate in certification schemes is to achieve higher or more stable coffee prices. In 2007, conventional coffee prices were good. Still, many farmers also judge current conventional and certified coffee prices insufficient for living and farm expenditures. Other farmers differentiate more and point out that they received a good price but that their yields were too low to make a living of it. In the research region, farm-gate prices for certified coffee were not always higher than for conventional coffee. In part this is due to the timing farmers chose for the final settlement of the bill. Conventional farmers received average farm-gate prices between 1.08US$/kg and 2.48US$/kg green coffee, the average farm-gate price for organic coffee varied from 1.34US$/kg to 2.55US$/kg. Between the Organic-Fairtrade certified cooperatives, farm-gate prices varied even more and ranged from 0.625US$/kg to 2.83US$/kg. Apart from the cooperative, the coffee prices also depend on the day farmers decided to liquidate their coffee and on the chosen marketing channel. In most cases farmers could not sell all their harvest at the highest prices but also sell to intermediaries (see Section 5.3.2).

For certified cooperatives high conventional coffee prices are a threat because farmers increase sales to conventional market channels. Certified cooperatives demand a certain coffee quality from their members which requires a higher labor input than conventional coffee. This either decreases leisure time or increases production costs when labor is hired. When conventional prices are low, farmers can clearly see the benefits of investing additional labor due to the higher premiums for certified coffee. With increasing conventional prices, the price differentials for certified coffees shrink. Farmers start selling their coffee in mainstream markets with lower quality requirements, thus reducing costs or work load. Because cooperatives often make their contracts with importers or roasters before or during harvesting time, the mainstream sales of farmers affect the cooperatives' ability to meet its contracts.

While the Fairtrade certification guarantees a minimum price to protect farmers against volatile coffee prices, many members of Fairtrade certified cooperatives could not connect anything else to the concept than its name. The consequences are twofold. First, farmers continue to be unwilling to invest on their farm because they fear future price drops. Second, when farmers do not know the price they are supposed to get, they cannot exercise control over the cooperative management on whether the received farm-gate price is justified. It is not easy to identify the reasons for this lack of understanding of Fairtrade. Little education may make it difficult to understand and remember the concept and some farmers also stated that they do not care much about the cooperative's explanations. But cooperatives may also present the information in a way that farmers face difficulties to follow as this reduces members' control over their activities. In some cooperatives, farmers do not dare to raise questions as they have been treated badly and their questions remained unanswered. Other farmers simply accept whatever the cooperative does. Often this attitude is not based on trust, but on a feeling of being powerless and dependent. However, here are differences between cooperatives and positive examples can also be reported, irrespective of the certification status.

5.3.2. Seasonal calendars

Together with the farmers seasonal calendars were elaborated with respect to working activities, the income, food and health situation. The calendars have been constructed by a farmer's group in each cooperative. Then, the information of each certification group has been summarized in one graph according to the explanations of the farmers. In the seasonal calendars, the time period during which an activity takes place, is slightly flexible, as it depends on the cooperative (e.g., the date credits are granted), the weather (e.g., the beginning of periods with increased sickness due to weather changes, the timing for labor peaks), and the altitude which affects for example the starting and end point of the coffee harvest. The seasonal calendars follow the coffee year in Nicaragua which starts in March and ends with the last coffee harvest in February (Figure 5.2).

In **coffee**, the harvest of coffee is usually between October and January. This is the highest labor peak within the year for all farmer groups, i.e. for conventional and certified coffee producers. Yet, in the course of the coffee year, there are several minor work peaks in coffee which are related to maintenance (pruning, fertilization) and weeding activities.

Not all farmers crop **maize and beans** but if they do, both are often intercropped or grown in two rotations within one year. Usually, it is the farmers living in 'lower' regions (between 800m a.s.l. and 1100m.a.s.l.) who crop maize and beans in addition to coffee. As maize and beans do not grow in higher altitudes, for those farmers living above 1100m a.s.l., transaction costs for cropping become too high if having to commute frequently to

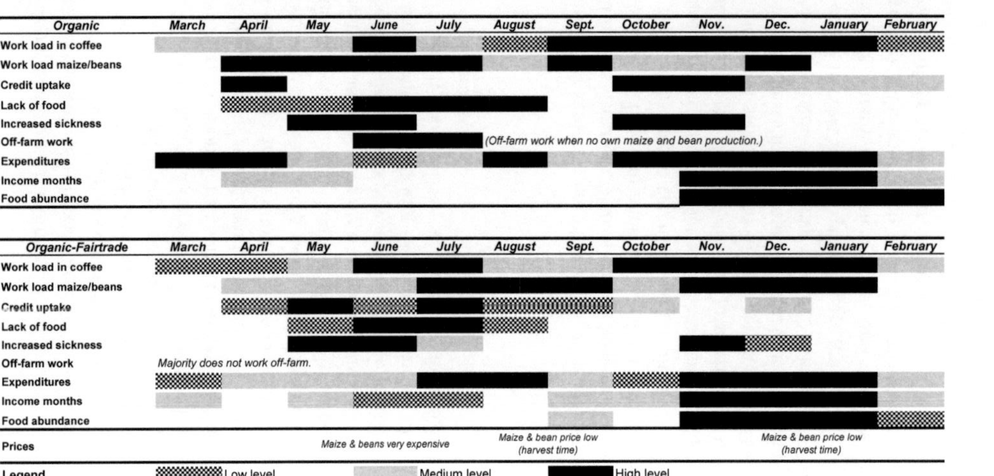

Figure 5.2.: Seasonal calendars of conventional, organic, and Organic-Fairtrade certified coffee farmers

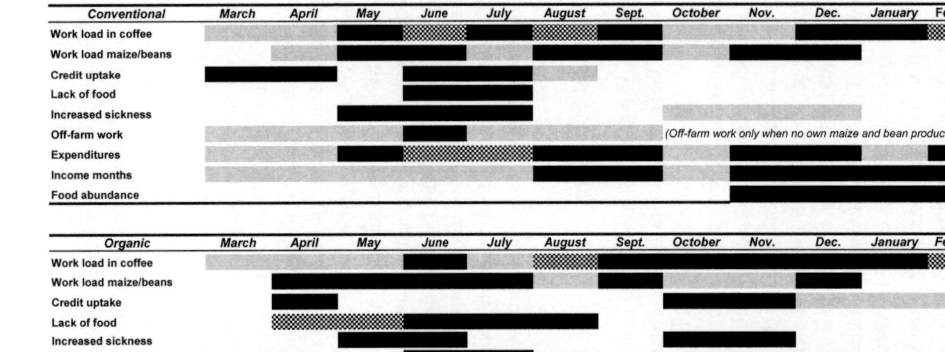

Source: Own illustration.

lower areas. The labor peaks in coffee coincidence with labor peaks or periods with high work loads in the maize and bean production. Farmers mentioned that there is sometimes a conflict of interest between the crops and that the work burden becomes extremely high. They employ off-farm labor if family labor is insufficient or have to neglect one of their crops. To increase yields and fight pests and diseases, even the organic certified coffee farmers usually apply chemical inputs in the maize and bean production. Only farmers of one Organic-Fairtrade certified cooperative could not do so as the organic standard, their coffee was certified with, also prohibited the use of chemical inputs in the whole farm and not only in the coffee production.

The time when farmers **take credits** usually depends in part on their cooperative. Some have certain months when they start issuing credits, usually after they finished the coffee sales, somewhere around April. This is also the time by which the farmers have spent most of their income obtained from the coffee harvest and start being in need of money. In May, the weather changes from dry to rainy season, in the high altitudes this shift leased frequently to **illness**. Flues, colds, coughing, and influenza are common. From May until end of July, **maize and bean prices rise** extremely. Even when farmers plant maize and beans, they usually suffer food shortages or reduce food diversity and quality, because they do not produce enough for the whole year. The high prices fall into the period when the stored food (either from own production or purchased in bulk during coffee harvest) is low. Conventional and certified farmers thus face at least two months of **food shortage** – the worst months being June and July, as they have to cut back in food consumption due to the high prices. Diets shift from maize and beans to plantain as it is intercropped with coffee. *"We only eat plantain with salt – sometimes for weeks"* (organic producer, July 2007). Certified farmers indicated longer periods with food shortage although not as such high levels as in June and July. These months coincide with the increased health problems due to the weather shift and the time when work load in maize, bean and coffee production is high – thus this time is a very difficult one for the farming families to cope with. When farmers have no own maize and bean production, either the husband but sometimes also the women search **work outside their farm** – the men on other farms or in the surrounding villages, the women look for domestic work in villages and cities in order to improve their financial situation. *"Those who work [on other farms] do it out of necessity. When there are obligations, when there are children[ ], when the child does not have milk, has no sugar, has no soap to wash the clothes"* (Organic-Fairtrade producer, April 2008). The conventional and organic farmers both indicated that sometimes they have to neglect their own farm in order to get immediate cash from work. The three farmer groups from the Organic-Fairtrade certified cooperatives indicated that the majority of them do not work off-farm. Another opportunity to obtain money is selling the coffee in advance to intermediary traders. The intermediaries offer cash against a fixed coffee quantity which has to be delivered at harvest time. Farmers rely on them only in emergencies as the offered farm-gate coffee prices are very low – between 15% and 25% of the coffee price farmers could obtain at the coffee harvest.

The difficult time for farmers ends latest in November with the onset of the coffee harvest. Those farmers who cropped maize and beans have a first **income** in August and September. From November until January are the main income months for all coffee farmers. Although the certified cooperatives usually pay their farmers later (between March and June), farmers sell to other market channels in order to have immediate cash for covering their financial needs. During that time, **food is abundant** and **expenditure** rise to cover the harvesting costs of coffee. The first income flows allow spending of money for food, health, clothes, and education. There was no difference between the conventional and certified farmers in this regard.

Summarizing, the graphical illustration of the seasonal calendar visualizes that the months of hardship range for conventional and certified farmers from May to July. During this time, farmers face labor peaks, lack of food and increased health problems. The months from November until January are months of increased prosperity as through the coffee sales income is generated and invested in household consumption.

5.3.3. Food consumption of conventional, organic and Organic-Fairtrade certified coffee households

This section focuses on the food consumption patterns of conventional, organic and Organic-Fairtrade certified coffee households. The quantitative household data indicates that on average, nearly 60% of farmers crop their own maize and around 65% their own beans. Organic certified coffee farmers take the lead, followed by the conventional farmers and Organic-Fairtrade have the lowest shares, yet very close to the other two groups (Table 5.1). Farmers buy either daily the needed maize and bean quantity or they buy less frequently than once a month. Among the Organic-Fairtrade certified farm households, it is also common to buy monthly their maize and bean supplies while the conventional farmers have to buy their staple food more frequently. The buying patterns may be influenced by poverty or remoteness. The poorest farmers can only afford to buy small quantities and those farmers living further away from the next shop will buy less frequently as transaction costs are high. Among the organic coffee farmers, 75% also grow their own beans, around 65% their own maize. Yet, the seasonal calendars as well as the semi-structured interviews indicated that food shortages are common among the organic farmers. This indicates that despite diversifying their agricultural production, organic farmers are not able to produce sufficient food for feeding their family throughout the year. It may relate to the trade-offs between coffee and staple crop productions due to labor constraints indicated in the seasonal calendars. Yield levels and efficiency of maize and bean production have not been further investigated, yet further research should also address these aspects especially in comparison to coffee production and identify where it would be optimal to invest resources.

Table 5.1.: Own maize and bean production and purchasing patterns of conventional, organic and Organic-Fairtrade certified farmers

	Conventional in %	Organic in %	Organic-Fairtrade in %	Total in %
Maize[m]				
Own production	55.9	65.7	52.4	58.4
Daily	10.3	10.2	11.4	10.7
Twice a week	2.9	0.0	0.0	0.7
Weekly	5.9	1.9	6.7	4.6
Every fortnight	5.9	0.0	2.9	2.5
Monthly	5.9	7.4	14.3	9.6
Less than monthly	13.2	14.8	12.4	13.5
Total	100.0	100.0	100.0	100.0
Beans[b]				
Own production	60.3	74.1	57.1	64.4
Daily	2.9	12.0	8.6	8.5
Weekly	5.9	4.6	4.8	5.0
Every fortnight	8.8	0.9	1.0	2.8
Monthly	8.8	3.7	15.2	9.3
Less than monthly	13.2	4.6	13.3	10.0
Total	100.0	100.0	100.0	100.0
N	281			

[m] Pearson $\chi^2 = 21.0932$, Pr $= 0.049$; [b] Pearson $\chi^2 = 30.7348$, Pr $= 0.001$
Source: Own data.

Table 5.2 shows the food consumption patterns of food items considered as 'luxury' food items by smallholders, like meat and cheese, as well as 'lower' quality food which is pure maize tortilla and plantain and the availability of stored food.

The consumption referred to the times the food item was eaten in the last seven days before the interview, given there were no special festivities. The cheapest of the 'luxury' food items, cheese, was most frequently consumed, followed by chicken. Summed up, the conventional farm households consumed on average 5.4 times a 'luxury' food item, organic households 5.0 times and Organic-Fairtrade 5.1 times. There was not much plain tortilla or plantain consumed but organic certified households consumed it more than Organic-Fairtrade households. Organic-Fairtrade had less rice per capita stored than the other two groups, while organic farmers stored most beans and maize. Their per capita availability of stored maize was significantly higher than in the other two groups. At the same time, the

Table 5.2.: Overview of food consumption and stored staple crops of conventional, organic and Organic-Fairtrade certified farm households

	Conventional mean (sd)	Organic mean (sd)	Organic-Fairtrade mean (sd)
Number of meals eaten in the past two days	6.0 (0.0)	5.9 (0.4)	6.0 (0.2)
Times chicken served in last 7 days	1.2 (1.4)	1.3 (1.0)	1.4 (1.3)
Times beef served in last 7 days	0.4 (0.8)	0.3 (0.5)	0.5 (0.9)
Times pork served in last 7 days	$0.4\ (0.7)^a$	$0.8\ (1.3)^a$	0.4 (0.7)
Times cheese served in last 7 days	3.4 (2.6)	2.6 (2.5)	2.8 (2.5)
Times only tortilla/plantain eaten in last 7 days	0.1 (0.6)	0.4 (1.3)	0.3 (1.1)
Days rice will last	$16.4\ (31.3)^a$	$17.3\ (24.5)^b$	$8.6\ (14.7)^{ab}$
Days beans will last	52.0 (67.3)	76.8 (109.7)	46.7 (69.6)
Days maize will last	$28.4\ (42.4)^b$	$79.8\ (114.4)^{ba}$	$42.5\ (66.8)^a$

Note: Superscript letters indicate a significant difference between two groups marked by the same letter.
[a] indicates a significant difference at $p < 0.05$.
[b] indicates a significant difference at $p < 0.01$.
Variables were non-normally distributed. Kruskal-Wallis tests were used to test for statistical differences, followed by pairwise comparisons based on the Mann-Whitney post hoc test adjusted by the Bonferroni correction factor (Field, 2005).

Source: Own data.

organic producers were in tendency those which indicated more insufficient food supplies. This may suggest that there is a larger spread between very poor and better-off households among the organic producers than in the other two groups. As stated above, also those coffee farming households which grow maize and beans mentioned shortages of food. Yet, their periods of food shortages are usually shorter than of those farmers who do not grow staple food crops. While in the periods of food shortages farmer cut down on consumption of animal products, maize and beans, they usually do not have to skip complete meals. During times of food shortages they still consume plantain, often only fried with salt, as all farmers intercrop plantain in their coffee plantations. This is the reason why in the large majority of farmers indicated that they had sufficient food available (Table 5.3). There is the tendency that slightly more organic coffee farming households had insufficient food supplies. There were also very few farmers who indicated larger periods of food insecurity in the past year. The conventional farmers have indicated that they did not experience any food insecurity in the past year which is a slight contradiction to Table 5.3. More organic than Organic-Fairtrade certified coffee producing households indicated food insufficiency (Table 5.4).

Table 5.3.: Food availability of conventional, organic and Organic-Fairtrade certified farmers in the past 30 days before the interview

Any days with insufficient food supply?

	Conventional in %	Organic in %	Organic-Fairtrade in %	Total in %
Yes	1.5	4.6	1.9	2.8
No	98.5	95.4	98.1	97.2
Total	100.0	100.0	100.0	100.0
N	281			

Note: Pearson $\chi^2 = 2.0436$, Pr $= 0.360$
Source: Own data.

When interpreting these results two aspects must be considered. First, data may be affected by selection bias. Second, data collection stretched for a period of four months (May-August) which included also the months of food shortages. The time when the interview took place may have also affected the result.

Table 5.4.: Number of food insecure days of organic and Organic-Fairtrade certified farmers in the last year

	Organic	Organic-Fairtrade	Total
Less than 10 days	5	2	7
Less than 30 but more than 10	5	4	9
Less than 180 but more than 30	1	0	1
More than 180 days	1	0	1
Total	12	6	18
N	18		

Note: Pearson $\chi^2 = 1.5714$, Pr $= 0.666$
Source: Own data.

5.3.4. The vicious cycle of indebtedness

Around three months after the coffee harvest, farmers face the hardest time of the year because by then, the income from coffee sales is spent, personal food stocks are depleted and food prices are high (cf. Section 5.3.2). Apart from reducing food consumption, farmers apply two other strategies. One strategy is to work as day-laborer. The other strategy is to obtain a credit from a cooperative or microfinance institution. Access to bank credits is difficult for small-scale farmers since banks require collateral and most farmers do not have legal land titles or sufficient animal stocks. Access to credit is very important for farmers and is their prime motivation for joining their cooperative. Most cooperatives offer credits but the issued amounts are often insufficient and usually no long-term credits are available. When credit necessities are higher than what is approved by their cooperative, farmers request credits from other organizations like local microfinance institutions or informal money lenders. If the credits are pursued from several organizations, it happens that the borrowed amount is higher than the payback capacity of farmers.

In many cases the credit is used for immediate consumption needs, like food or medicine and only partially invested in the farm. Thus, only some farm maintenance is done, fertilization is insufficient and long-term investments like replanting of old coffee trees are very limited. Consequently, the state of the plantation at best remains the same, harvested yields stay low, leading again to a low income despite the obtained premiums for the organic and Organic-Fairtrade certified coffees. Especially if debts need to be paid back, little of the yearly income is left for household consumption and farm management. As one producer explained: *"Sometimes we borrow a bit more money for the coffee harvest, and thus, sometimes we do not get anything in the final settlement, sometimes we continue to be in debt"* (Organic-Fairtrade certified producer, April 2008). So farmers have to apply for a new credit soon after having finished the coffee. They often are thus trapped in a vicious cycle of indebtedness (Figure 5.3). A mayor contributing factor to the vicious cycle is that most farmers have no overview of their production and living costs or the amount of money needed for debt payment. Although organic certification requires book-keeping this is often not well done, especially balance sheets are not filled. The unawareness of costs may be explained by farmers' low educational levels. In Nicaragua, around 33% of population over 15 years is unable to read or write (World Bank, 2007). As most small-scale farmers are financial illiterate, they often sign whatever they are told to get the money. Even cooperatives credits have minimum annual nominal interest rates of 18%. Private microfinance institutions in the region have annual nominal interest rates of up to 36%. The effective interest rates are in both cases much higher. Very often farmers are not aware of the effective interest rate because hidden costs like administration costs, obligatory savings, and risks of exchange rate variations are also added, but are not always explicitly listed in the contracts given to farmers. Even given the case this information is provided most farmers are often not able to calculate the effective interest rate they are

Figure 5.3.: The vicious credit – yield cycle

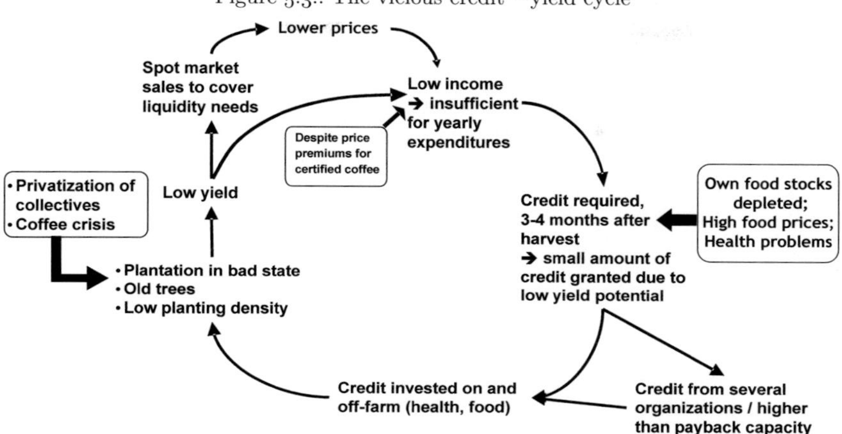

Source: Own illustration.

paying. Like that, farmers are often unaware by the amount of money required to cancel the debt.

Given the low yield and thus low income levels, it is not surprising that, like the conventional farmers, the majority of certified farmers consider themselves to be poor. *"The majority of us do not have money; the majority of us are poor"* (Organic-fair trade producer, April 2008). Despite participating in certification schemes, farmers feel trapped in their poverty: *"Because we are all poor, we cannot escape the misery"* (Organic-Fairtrade producer, April 2008).

Regarding producers' poverty, a cooperative manager pointed out that even with doubling the Fairtrade minimum price, cooperative members would still continue to live in poverty as yields are too low for making a sustainable living. For a good sustainable organic farm management, including some replanting of coffee trees, the cooperative estimates production costs to be around 3.02US$/kg green coffee at current yield levels of around 390kg/ha.[3] The cooperative's farm-gate prices for Organic-Fairtrade coffee were highest among all cooperatives. Still, average yield levels need to increase around 50% to reach the break even point, or the farm-gate coffee prices need to increase by further 9% respectively for breaking. Then, only the farm would be optimally managed but not yet an income from it generated.

---

[3] Further 0.66US$/kg for the processing costs and cooperative fee need to be added to the costs. The cooperative reached on average a green coffee price of 3.37US$/kg (Free on Truck-FOT), excluding the social premium.

5.4. Conclusions

Results show that in northern Nicaragua coffee yields are usually low due to limited maintenance activities and badly managed plantations irrespective of the certification. Certification schemes offer higher farm-gate prices, but differences between cooperatives exist. In general, farmers with Organic-Fairtrade certification received higher farm-gate prices than farmers with only the organic or without certification. At given yield levels, farm-gate prices are not sufficient to offset production costs for optimal organic farm management. Producers' income from conventional and certified coffee is insufficient for living and farm expenditures. Necessary financial needs are covered by credits, which, when combined with financial illiteracy, result in a vicious cycle of indebtedness.

Conventional and certified coffee farmers in the region face food shortages but do not have to skip whole meals. There is the trend that certified coffee farmers are slightly more affected by insufficient food intake than conventional farmers. Conventional face similar periods of hardship and increased prosperity like the certified farmers, no major differences which could indicated an increased well-being of organic or Organic-Fairtrade farmers could be identified.

The qualitative research results indicate that organic and Fairtrade coffee certifications in northern Nicaragua tend to have a low impact on food security. It is likely that the main causes of continuing food insecurity and poverty among smallholder coffee growers in Nicaragua are not the lack of market access or unfair trading conditions. Reasons for poverty and food insecurity seem rather low yield levels, low educational levels and farmers' undeveloped entrepreneurial skills as well as financial constraints, credit constraints and indebtedness. Certification schemes do not address or solve these problems and thus can only be part of a development policy for poor, small-scale farmers in Nicaragua.

The vicious cycle of indebtedness of smallholder coffee producers which was discovered in this research merits further investigation which addresses in detail the access to credits, nominal and real interest rates of credits, income and expenditure patterns to identify possible policy intervention areas and ways to alleviate the financial constraints of farmers. Regarding the development target of poverty alleviation and food security, policy makers, NGOs and international donors need to specifically address the causes of poverty. In the case of Nicaragua's small-scale coffee farmers, policies need to enhance coffee yields, for example through research in improved and adapted varieties and facilitating the access to fertilizer. Business training for farmers, for example through agricultural extension services, is necessary and must also enhance financial literacy.

Acknowledgements
The funding for field research by the 'Advisory Service on Agricultural Research for Development (BEAF)' of the 'Deutsche Gesellschaft für Technische Zusammenarbeit (GTZ)' and the logistic support by the 'International Center for Tropical Agriculture (CIAT)' in Colombia and Nicaragua, especially by Dr. Axel Schmidt, is gratefully acknowledged.

5.5. Bibliography

Arnould, E. J., Plastina, A., Ball, D., 2007. Market disintermediation and producer value capture: The case of fair trade coffee in Nicaragua, Peru, and Guatemala. Advances in International Management 20, 319–340.

Bacon, C., 2005. Confronting the coffee crisis: Can fair trade, organic and specialty coffees reduce small-scale farmer vulnerability in northern Nicaragua? World Development 33 (3), 497–511.

Bacon, C., Ernesto Mendez, V., Gomez, M. E. F., Stuart, D., Flores, S. R. D., 2008. Are sustainable coffee certifications enough to secure farmer livelihoods? The Millennium Development Goals and Nicaragua's fair trade cooperatives. Globalizations 5 (2), 259–274.

Bray, D. B., Sánchez, J. L. P., Murphy, E. C., 2002. Social dimensions of organic coffee production in Mexico: Lessons for eco-labeling initiatives. Society & Natural Resources 15 (5), 429–446.

Cashin, P., McDermott, C. J., Scott, A., 2002. Booms and slumps in world commodity prices. Journal of Development Economics 69 (1), 277–296.

Daviron, B., Ponte, S., 2005. The coffee paradox. Global markets, commodity trade and the elusive promise of development. Zed Books, London and New York.

Field, A., 2005. Discovering statistics using SPSS, 2nd Edition. SAGE Publications, London, Thousand Oaks, New Delhi.

Fitter, R., Kaplinsky, R., 2001. Who gains from product rents as the coffee market becomes more differentiated? A value-chain analysis. IDS Bulletin 32 (3), 69–82.

FLO, 2007. Annual report 2007. An inspiration for change. Fair Trade Labelling Organizations International (FLO), Bonn, accessed 15 December 2010.
URL http://www.fairtrade.at/pics/infomaterial/FLO_AR2007.pdf.

FLO, 2008. Fairtrade standards for coffee for small farmers' organizations. Fair Trade Labelling Organizations International (FLO), Bonn.

Giovannucci, D., Villalobos, A., 2007. The state of organic coffee: 2007 US update. Sustainable Markets Intelligence Center (CIMS), San Jose, accessed 15 December 2010.
URL http://orgprints.org/12856/1/State_of_Organic_Coffee_2007_US_update.pdf

Gresser, C., Tickel, S., 2002. Mugged: Poverty in your cup. Oxfam International; Make Trade Fair, Washington D.C., accessed 21 November 2006.
URL `http://www.oxfamamerica.org/newsandpublications/publications/research_reports/mugged`.

ICO, 2004. Lessons from the world coffee crisis: A serious problem for sustainable development. ED 1922/04 International Coffee Organization (ICO), London, accessed 3 February 2011.
URL `http://www.ico.org/documents/ed1922e.pdf`

IFOAM, 2006. Organic agriculture and food security. International Federation of Organic Agriculture Movements (IFOAM), Bonn, accessed 20 July 2006.
URL `http://www.ifoam.org/organic_facts/food/pdfs/Food_Security_Leaflet.pdf`.

IFOAM, 2009a. Definition of organic agriculture. International Federation of Organic Agriculture Movements (IFOAM), Bonn, accessed 29 March 2010.
URL `http://www.ifoam.org/growing_organic/definitions/doa/index.html`.

IFOAM, 2009b. The principles of organic agriculture. International Federation of Organic Agriculture Movements (IFOAM), Bonn, accessed 15 December 2010.
URL `http://www.ifoam.org/about_ifoam/principles/index.html`.

Lewin, B., Giovannucci, D., Varangis, P., 2004. Coffee markets: New paradigms in global supply and demand. Agriculture and Rural Development Discussion Paper No. 3. World Bank, Washington D.C.

Linton, A., 2008. A niche for sustainability? Fair labor and environmentally sound practices in the specialty coffee industry. Globalizations 5 (2), 231–245.

Murray, D. L., Raynolds, L. T., 2006. The future of fair trade coffee: Dilemmas facing Latin America's small-scale producers. Development in Practice 16 (2), 179–191.

Mutersbaugh, T., 2002. The number is the beast: A political economy of organic-coffee certification and producer unionism. Environment and Planning A 34, 1165–1184.

Osterhaus, A. (Ed.), 2006. Business unusual. Successes and challenges of fair trade. FLO (Fairtrade Labeling Organizations International), IFAT (International Fair Trade Association), NEWS! (Network of European Worldshops) and EFTA (European Fair Trade Association), Newcastle-upon-Tyne.

Raynolds, L. T., Murray, D., Taylor, P. L., 2004. Fair trade coffee: Building producer capacity via global networks. Journal of International Development 16 (8), 1109–1121.

Rice, R. A., 2001. Noble goals and challenging terrain: Organic and fair trade coffee movements in the global marketplace. Journal of Agricultural and Environmental Ethics 14 (1), 39–66.

Scoones, I., 1998. Sustainable rural livelihoods-framework for analysis. IDS Working Paper No. 72. Institute of Development Studies (IDS), Brighton.

Vakis, R., Kruger, D., Mason, A. D., 2004. Shocks and Coffee: Lessons from Nicaragua. Social Protection Discussion Paper Series No.0415. World Bank, Washington, DC.

Varangis, P., Siegel, P., Giovannucci, D., Lewin, B., 2003. Dealing with the coffee crisis in Central America. Impacts and strategies. Policy Research Working Paper No. 2993. World Bank, Washington D.C.

Varughese, G., Ostrom, E., 2001. The contested role of heterogeneity in collective action: Some evidence from community forestry in Nepal. World Development 29 (5), 747–765.

WFTO, 2009. What is fair trade? World Fair Trade Organization WFTO, Culemborg, accessed 22 January 2011.
URL `http://www.wfto.com/index.php?option=com_content&task=view&id=1082&Itemid=11`.

WFTO, 2010. Charter of Fair Trade Principles. Introduction. World Fair Trade Organization WFTO, Culemborg, accessed 22 January 2011.
URL `http://www.wfto.com/index.php?option=com_content&task=view&id=1082&Itemid=11`.

Wollni, M., Zeller, M., 2007. Do farmers benefit from participating in specialty markets and cooperatives? The case of coffee marketing in Costa Rica. Agricultural Economics 37 (2-3), 243–248.

World Bank, 2007. World Development Report 2008: Agriculture for Development. World Bank, Washington D.C.

— Chapter 6 —

Organic and Fairtrade Coffee Production: The Role of Cooperative Business Models for the Success of Smallholder Coffee Certification in Nicaragua

Tina Beuchelt and Manfred Zeller

This chapter has been accepted in December 2011 for publication as article by the journal Renewable Agriculture and Food Systems.[1]

Abstract

Supported by governments, NGOs, and donors, coffee farmer organizations obtain organic and Fairtrade certifications to upgrade their coffee and thus to increase returns for their smallholder members. Whether this upgrading strategy fits into the business model of the cooperative is often not considered; the success of certification strategies varies. As research on this topic is scarce, we analyze the business models, upgrading strategies and strengths, weaknesses, opportunities, and threats (SWOTs) of seven conventional, organic and Organic-Fairtrade certified coffee cooperatives and link these to member's coffee gross margins. We use data from over 100 in-depth qualitative interviews and a household survey of 327 cooperative members in northern Nicaragua. We find that cooperatives often apply the same upgrading strategies despite very different business models and SWOTs, leading to different levels of success. There are also many commonalities of SWOTs among cooperatives, such as weak infrastructure, management problems or limited credit access, which can be addressed by governmental policies. The qualitative comparison of coffee gross margins among the cooperatives shows no clear income effect from participating in certified coffee production but indicates dependence on the business model and upgrading strategies of the cooperatives. Upgrading through certification seems only successful with a suitable business model and other upgrading strategies. Policies should focus on the institutional framework of cooperatives by offering strategic advice, credit access, and external auditing of cooperatives. The national coffee sector strategy could be supported by establishing a national coffee federation or institute as in other countries.

---

[1]A revised version of this article is published in: Beuchelt, T.D. and M. Zeller. The role of cooperative business models for the success of smallholder coffee certification in Nicaragua: A comparison of conventional, organic and Organic-Fairtrade certified cooperatives. Renewable Agriculture and Food Systems, Available on CJO doi:10.1017/S1742170512000087. Copyright by the Cambridge University Press. This chapter is reprinted with permission.

Keywords:
Business model; certification; cooperative; Fairtrade; gross margin; organic; SWOT analysis; upgrading strategy

Abbreviations:
CONV   Conventional Cooperative
IFOAM   International Federation of Organic Agriculture Movements
FLO   Fair Trade Labelling Organization
NGO   Non-Governmental Organization
ORG   Organic Certified Cooperative
OFT   Organic-Fairtrade Certified Cooperative
SWOT   Strengths, Weaknesses, Opportunities, and Threats

6.1. Introduction

Food products and production processes following environmental, social, quality and safety standards have become increasingly popular among traders, retailers, and consumers in industrialized countries (Asfaw et al., 2010; Basu et al., 2003; Codron et al., 2006; Swinnen and Maertens, 2007; Ponte and Gibbon, 2005; Rigby and Cáceres, 2001). In the last decade, organic certified products have moved from niche to mass markets in industrialized countries; a similar trend is likely to emerge for ethical and fairly traded products (Codron et al., 2006). The shift in consumer awareness and demand, including the willingness to pay higher prices for certified and/or high quality products, has also been noticed by governments in developing countries, NGOs, and international donors (Bolwig et al., 2009; Pelupessy and Díaz, 2008). To address the deteriorating terms of trade of agricultural commodities (Kaplinsky, 2000), price volatility (Cashin et al., 2002), and poverty among agricultural smallholders in developing countries, donors and policy-makers recommend to upgrade agricultural production and to enter high-value markets. Upgrading strategies are supposed to increase the captured value, the performance, the competitiveness or the position of farmers and farmer organizations/cooperatives (Humphrey and Schmitz, 2002; Markelova et al., 2009; Riisgaard et al., 2010). Coffee producers and their cooperatives can upgrade by offering products with higher quality or additional environmental or social certification, or by processing and exporting the goods themselves (Bolwig et al., 2009; Kilian et al., 2006; Linton, 2008; Willer and Yussefi, 2007). The food and agricultural commodity value chains shifted from domestically oriented, state-controlled systems to globally integrated food supply chains with private governance, leading to new forms of vertical coordination (Swinnen and Maertens, 2007). Cooperatives can make an important contribution by linking farmers to the new, vertically more integrated, food supply chains and specialty, high value markets (Wollni et al., 2010). Each cooperative thereby follows a certain business model, which refers to the way the cooperative is run, its internal organization, values, and services offered like credit or marketing services.

Farmer organizations, cooperatives or companies are under pressure to increase their competitiveness and improve their performance, especially as new producers enter the market (Humphrey and Schmitz, 2002). In the coffee sector, smallholder producers and cooperatives have to defend themselves against low-cost producers like Brazil or Vietnam. The sector is also characterized by frequent oversupplies, production shocks, and high price volatility (Cashin et al., 2002). The linking of smallholder farmers and their cooperatives to high value (coffee) markets offers potentially great rewards, but there are also considerable challenges (Kaganzi et al., 2009). Policy-makers and donors often present the same upgrading strategies to coffee cooperatives without considering their individual business models, thereby ignoring the heterogeneity of farmer organizations, their internal structures, existing management capacities or available infrastructure.

A typically suggested upgrade for coffee producers is the participation in certification schemes like the organic or Fairtrade certification. Both certifications aim at increasing the welfare of smallholder producers through improving production and marketing processes (IFOAM, 2006a,b; Kilian et al., 2006; Linton, 2008; Wills, 2006). The focus of the organic certification is more on environmentally friendly production processes while the Fairtrade certification focuses on social standards and fair marketing and trading conditions. We will not further describe the standards of the organic and Fairtrade certification as this has been done by many other studies (Barrett et al., 2001; Osterhaus, 2006; Raynolds, 2000; Raynolds et al., 2007; Rigby and Cáceres, 2001). While the Fairtrade certification is only available to coffee producers organized in a group or cooperative, the organic certification is also issued to individual producers. Since organic certification costs are very high, smallholders can usually only afford certification when they participate in a farmer group that splits the cost among members. Several research studies have indicated that the organic and Fairtrade certification can result in price premiums and economic benefits at the producer level in Latin America (Bacon, 2005; Raynolds et al., 2004; Wollni and Zeller, 2007), while other research has found little economic benefits for farmers (Kilian et al., 2006; Mutersbaugh, 2002; Neilson, 2008; Philpott et al., 2007). In addition to providing possible economic benefits, organic and Fairtrade certifications are known to improve education, health and infrastructure (Arnould et al., 2009; Bacon et al., 2008), and to increase social organization and capacity building (Bray et al., 2002; Raynolds et al., 2004; Taylor et al., 2005).

So far, the existing literature either covers the determinants for successful collective action, group formation, farmer organizations, and resulting advantages and disadvantages (Hellin et al., 2009; Kaganzi et al., 2009; Markelova et al., 2009; Pickard, 1970; Thorp et al., 2005) or it addresses the socio-economic effects of coffee certification on smallholders with a focus on farm, household- and community-levels (Bacon et al., 2008; Barham et al., 2011; Valkila, 2009). The studies only marginally touch the business model and upgrading strategies of farmer organizations which together affect the outcome of certification schemes. The fit of upgrading strategies to the business model of a farmer organization as well as the determinants for the success of upgrading through certification has so far been neglected by research. This may also be one explanation for the contradictory findings regarding economic benefits in the above cited studies. Additionally, most of the cited studies on certification effects have neither randomly sampled cooperatives/farmer organizations nor analyzed several cooperatives or certification standards simultaneously. Arnould et al. (2009) worked with randomly sampled cooperatives but neglected to analyze the business models. Raynolds et al. (2004) systematically compare seven, non-randomly selected, Fairtrade certified producer cooperatives in Mexico, Guatemala and El Salvador, to see the characteristics that facilitate successful integration into Fairtrade networks and the resulting benefits. They find that successful Fairtrade participation depends on the political economic and marketing conditions, the social and ecological resources of producers, and the producer groups' internal organization and external links (Raynolds et al., 2004), yet they did not analyze the latter in depth. Only one certification scheme is targeted and as no comparison with cooperatively organized conventional producers was done, their study design cannot differentiate between the socio-economic effect of certification and the socio-economic effect of being organized in a cooperative or producer group.

Raynolds et al. (2004) assume that their above identified characteristics haven proven to be historically successful but are likely to be insufficient in the future given the growing competition in Fairtrade markets. Therefore, we focus specifically on the roles of the producer groups, their business models and upgrading strategies for the success of certification schemes. Our paper fills the identified research gaps through the analysis of cooperatives with two certification schemes, organic and Organic-Fairtrade, against a control group of also cooperatively organized but conventional coffee smallholders.

The following three research questions are analyzed:

1. What are the business models and upgrading strategies of conventional, organic and Organic-Fairtrade certified cooperatives?
2. What are the strengths, weaknesses, opportunities, and threats of conventional, organic and Organic-Fairtrade certified cooperatives?
3. How are the coffee gross margins linked to the organic and Organic-Fairtrade certification strategy, the cooperative's business model and its other upgrading strategies?

We collected over 100 qualitative in-depth interviews with conventional, organic and Organic-Fairtrade certified smallholder coffee producers, their representatives and staff of seven cooperatives in northern Nicaragua. Additionally, we conducted a two-stage random sample based household survey with 327 coffee producers of the seven cooperatives.

The article is structured in seven sections. The next section portrays the elements of the conceptual framework. The third section describes the sampling framework and the analytical methods. In the fourth section, the business models and upgrading strategies of the conventional, organic and Organic-Fairtrade certified cooperatives are described. This is followed by an analysis of the strengths, weaknesses, opportunities, and threats of these cooperatives in the fifth section. The results of the fourth and fifth section are linked to the coffee gross margins obtained by the cooperative members in the sixth section. The last section contains the conclusions and policy recommendations.

6.2. The conceptual framework: Business models and upgrading strategies of coffee cooperatives

It is increasingly recognized that the potential of smallholders to raise their agricultural incomes depends on their ability to participate and compete successfully in markets (Hellin et al., 2009; Markelova et al., 2009). Farmer organizations such as cooperatives can play an important role in connecting producers to markets. A market-oriented cooperative is characterized by a group of economically active individuals who engage in a joint undertaking with the objective of deriving benefits for themselves as members (Dülfer, 1995; Kramer, 2000). Over the past few decades, as cooperatives have become increasingly aware of market forces, their organizational structures have adapted, leading to a broader range of cooperative structures than before (Kramer, 2000). Each cooperative, like a company, follows an individual business model. In simple words, business models can be seen as *"stories that explain how enterprises work"* (Magretta, 2002, p 87). The business models of cooperatives can be classified according to micro-economic aspects like the main activity (credit/savings, production, consumption, service), the type of economic functions or services (supply/purchase, marketing, production, financing, administration), the number of functions and services, and the degree of organization (informal or formal) (Zerche et al., 1998). Service cooperatives can offer input supply, processing and marketing of outputs, financing, or provide technology and can also support their members in entering high-value markets (Deininger, 1995; Markelova et al., 2009). Especially in the agricultural sector of developing countries, cooperatives offer multiple services.

Shafer et al. (2005) define a business model as a description of the underlying logic of a company or cooperative and its strategic choices for creating and capturing value within a value chain. This implies that cooperatives create additional value by differentiating themselves from competitors through pursuing a strategy (Magretta, 2002). Upgrading strategies are mainly seen as strategies which improve the position of producers or cooperatives in the value chain through shifting to more (economically) rewarding functional positions, for example through the uptake of new activities previously performed by other chain actors, or through increasing the added value of production or improving returns (Bolwig et al., 2010; Gereffi, 1999; Giuliani et al., 2005; Humphrey and Schmitz, 2002). In a broader definition, upgrading can additionally include a change of activities which leads to reduced exposure to risk (Ponte and Ewert, 2009; Riisgaard et al., 2010). This definition embraces the possibility of downgrading. For small producers and entrepreneurs in developing countries, Riisgaard et al. (2010) identify seven possible upgrading strategies for smallholders and their cooperatives which can be grouped in three types according to (i) the improvement of the product, volume or production process, (ii) the change and/or the adding of functions in the chain, and (iii) the improvement of value chain coordination through horizontal and vertical contracts. Improvements of the product can be achieved through increasing quality, complying with certification, such as organic or Fairtrade, or by meeting food safety standards. Improvements of the produced volume can be obtained through increases in yield or area, whereas the production process can be improved through increasing efficiency or reducing negative externalities such as environmental pollution. The change of functions in the chain refers to either a) functional upgrading such as taking on new functions in the value chain like processing, exporting, roasting or providing services/inputs, or b) functional downgrading through stoppage of unprofitable activities and concentration on core activities. Vertical value chain coordination can refer to improving business ties with buyers through contracts instead of spot market sales, for example, cooperatives may contract directly with coffee roasters. Horizontal coordination can be agreements among producers or cooperatives to co-operate over marketing, services, and input provision.

The conceptual framework

To summarize, the conceptual framework builds on the cooperative, its business model and chosen upgrading strategies (Figure 6.1). We use the term business model for the cooperatives based on the above mentioned definitions, but limit the analysis to organizational structure, size, values, functions, services, and financial characteristics. We apply

Figure 6.1.: The influence of the cooperative's business model and upgrading strategies on members' coffee income

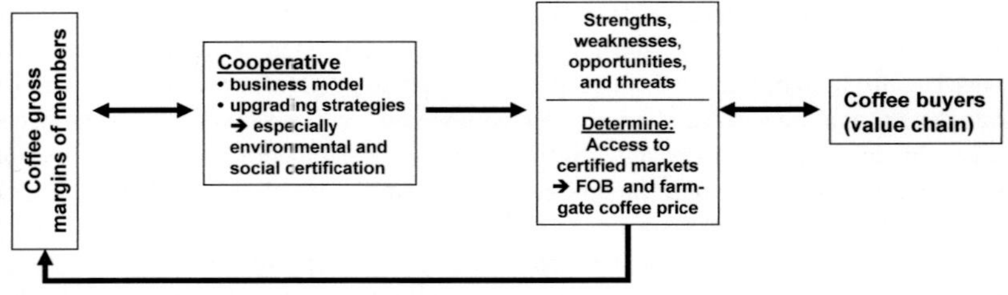

Source: Own illustration.

the classification of upgrading strategies according to the framework of Riisgaard et al. (2010). A focus is on the upgrading strategy of environmental and social certifications such as the organic and Organic-Fairtrade certification. The business model and upgrading strategies together determine the strengths and weaknesses of the cooperative. These are internal factors which affect the cooperative and its members. External factors emerge from the natural and business environment within which the cooperative operates. They constitute opportunities for development as well as threats. The internal and external factors determine the access to certified markets as well as the coffee prices the cooperative obtains in international markets. These affect the farm-gate prices the cooperative can offer and thus the coffee income possibilities of the cooperative's members. The coffee income is measured by the coffee gross margins. The market access and the coffee prices naturally depend also on the quality produced and the cooperative's coffee buyers such as exporters, importers, and roasters.

6.3. Sampling framework and methodology

The research was conducted in northern Nicaragua, in the departments Madriz and Nueva Segovia, on coffee farms situated between 900m and 1300m a.s.l. The research region was homogeneous in terms of living conditions and the socio-economic level of the smallholders, as well as the coffee growing characteristics found on their farms. Quantitative household data was collected in 2007. Cooperatives were selected randomly from a self-established list as no official list with all cooperatives in that region existed. We classified cooperatives according to their certification into conventional, organic and Organic-Fairtrade certified cooperatives. To select the cooperative members, either a simple random sampling or a two-stage random sampling was applied depending on the cooperative. Using a structured questionnaire, 327 cooperative members were surveyed with nearly equal shares of conventional, organic and Organic-Fairtrade producers. One conventional cooperative dropped out of the study near the end of the data collection period. The conventional cooperatives are called CONV-A (n=37) and CONV-B (n=31); the organic certified cooperatives ORG-B (n=32) and ORG-C (n=76); and the Organic-Fairtrade certified cooperatives OFT-D (n=34), OFT-E (n=36), and OFT-F (n=34). ORG-B is a subgroup of organic coffee producers of the conventional cooperative CONV-B.

In addition, qualitative data were collected with 48 key-person interviews of cooperative staff, exporters, roasters, and researchers, 33 semi-structured producer interviews, and 21 focus group discussions, with an average of nine participants in each group, in 2007 and 2008. The interviews with the cooperative staff were dealt with the business model of the cooperative, its upgrading strategies, coffee sales, its strengths, weaknesses, opportunities, and threats. All interviewees gave their informed consent prior to the quantitative and qualitative interviews.

We use an analytical method called 'Strengths, Weaknesses, Opportunities, and Threats' (SWOT) analysis. SWOT helps to provide a quick picture of an organization's current position and its short-term future (Fleisher and Bensoussan, 2003). This overview can be used to evaluate the current situation and make strategic decisions. This qualitative approach combines the internal strengths and weaknesses of a cooperative with the external influences of a business environment to identify the opportunities and threats faced by a cooperative (Dabbert and Braun, 2009). Being a descriptive model it does not offer explicit strategies (Fleisher and Bensoussan, 2003), but provides a transparent overview and reveals important problem areas (Rauch, 2007). We use the SWOT analysis to identify internal and external factors that affect the cooperatives and their implicit goal to increase their members' economic welfare. The upgrading strategy of 'certification' is compared with the 'business as usual' model of conventional cooperatives. The questions listed in Table 6.1 are the guiding questions used in the interviews with the cooperative staff. In addition, further key-persons like coffee exporters and NGO staff were also asked these questions to obtain a broader view of the coffee sector in Nicaragua.

Table 6.1.: Questions regarding cooperatives' strengths, weaknesses, opportunities, and threats

Strengths	Weaknesses
• Where do you and your buyers see the strengths of your cooperative? • Which activities of the cooperative are going well? • What have been your most notable achievements?	• What are limitations or problems the cooperative faces? • What relevant resources do you need?
Opportunities	**Threats**
• Where do you see the opportunities of your organization? • Is there an opportunity to demand better prices from suppliers?	• Where do you see the threats to your coop.? • Have there been any significant changes in the coffee industry or governmental regulations? • Are there any, or do you anticipate new competitors in your market?

In a next step, we link the cooperative business models, SWOTs, and upgrading strategies with the member's coffee gross margins. Gross margins are calculated per hectare of coffee. Even certified coffee producers usually sell to various coffee buyers outside their cooperatives, these sales are called 'off-sales'. We calculate therefore the revenue per hectare by summing up the coffee quantity sold in a market channel multiplied by the farm-gate coffee price obtained in that market channel; divided by the coffee area in hectares. Farm-gate coffee prices were discounted where necessary. In addition, we indicate the average price coffee producers received across all channels by dividing the total revenue by the total quantity sold. The variable production costs are deducted from the revenue to obtain the gross margin. The variable production costs include all input costs like organic or chemical fertilizer, pest, disease and weed treatments, transport costs, the costs occurred at harvest, and the costs for hired labor. The gross margins of the cooperative members are used to indicate the success of the certification scheme.

6.4. The cooperative business models and their upgrading strategies

The structures of cooperatives in Nicaragua are currently undergoing changes in response to a law passed in 2004 obliging cooperatives to register legally and to reorganize their structures according to the new regulations. This has led to a relatively homogenous organizational structure within the currently registered cooperatives. However, not all existing cooperatives finished this registration process, largely because they could not afford to pay for the obligatory training of cooperative members and staff all at once. The hierarchical organization of cooperatives distinguishes first-, second- and third-level cooperatives. A first-level cooperative is the lowest level where individual farmers are members. The organizational components of first-level cooperatives are the general assembly (either of members or elected delegates), the supervisory board which controls finances, and the administrative board or steering committee composed of the president, vice president, and treasurer. The steering committee is supported by several specialized committees, e.g. for education and cooperativism, credits, or commercialization. A minimum of five first-level cooperatives can form a second-level cooperative to use economies of scale in processing, marketing or exporting of products. Second-level cooperatives can then group together to form third-level cooperatives or federations. Many variations of this still exist because the law has not yet been fully implemented.

The values inherent to each cooperative are indicated in the written vision and mission statements of the cooperatives and their core content is summarized in Table 6.2. The statements represent the cooperative's objectives according to which members can judge the achievements of their cooperative. While there are many overlaps in the statements, there are also differences between the cooperatives. The conventional cooperatives have a stronger focus on financial services in their mission statement than the certified cooperatives – despite the fact that all cooperatives are multifunctional cooperatives. Except the Organic-Fairtrade certified cooperative OFT-D, all cooperatives emphasize providing

Table 6.2.: Summary of mission and vision statements of conventional, organic and Organic-Fairtrade certified cooperative business models

	CONV-A	CONV-B / ORG-B	ORG-C	OFT-D	OFT-E	OFT-F
Provide financial services	x	x				
Offer non-financial services and/or member support	x	x	x		x	x
Be organic farmer organization based on environmental & social responsibility				x		
Improve living standard/welfare of members	x	x	x		x	x
Strong participation of members						x
Produce and market high coffee quality				x		x
Sell coffee nationally and internationally				x	x	
Be a financially solid organization	x	x	x		x	x
Become (more) self-financed			x			x

non-financial services to their members and contributing to their members' socio-economic development. The Organic-Fairtrade certified cooperatives are more focused on producing high quality and finding a good market position for their coffee than the organic and conventional cooperatives. Yet, both organic certified cooperatives know that they already produce a high quality coffee and emphasized that aspect in the interviews. An important element in the vision of all cooperatives is to become financially both stable and independent as none of them are yet (see also Section 6.5). An exception is cooperative OFT-D which is so deeply indebted that it had to accept an external manager chosen by the second-level cooperative to which it owes the money. The cooperative OFT-F has a very strong mission statement with respect to the economic and social promotion of its members. The cooperative ORG-C has no written mission and vision, but the qualitative research revealed that the cooperative leaders and members have the cooperative values internalized in their daily life.

Table 6.3 provides an overview of the cooperatives' business models, their characteristics and services. The two conventional cooperatives focus on offering financial services such as provision of production and consumption credits as well as savings accounts. Cooperative CONV-A offers only few additional services and markets coffee through an alliance with an exporter. Cooperative CONV-B offers many additional services such as supplying

Table 6.3.: Overview of the characteristics and services of conventional, organic and Organic-Fairtrade certified cooperative business models

	CONV-A	CONV-B/ORG-B	ORG-C	OFT-D	OFT-E	OFT-F
Cooperative type	First-level	First-level, member of 2nd-level coop.	First-level	First-level, member of 2nd-level coop.	2nd-level, composed of 8 first-level coops.	2nd-level, composed of 6 first-level coops.
Year of foundation	1998	1994	1999/2000	1999	Restructured in 1992	1996
Total member size	338	9313	130	600	309	168
Type of members	Small sized agric. producers, mainly in coffee	Small/medium sized agric. & industrial producers, employees	Smallholder coffee producers	Smallholder coffee producers, some small agric. producers	Smallholder coffee producers	Smallholder coffee producers
Type of activity: Economic functions and services	Credit/savings & service: Financing, both input supply & limited marketing	Credit/savings & service: Financing, input supply, marketing (limited)	Service: Input supply, coffee marketing, credit (limited)	Service: Input supply, coffee marketing, credit (limited)	Service: Input supply, coffee marketing, credit (limited)	Service: Input supply, coffee marketing, credit (limited)
Marketing of coffee	Through alliance with exporter (does also processing)	Own dry mill but markets via exporter & 2nd-level coop.	Own marketing (via exporter who does also processing)	Own dry mill, own marketing and via 2nd level coop.	Dry mill contracted, own marketing, has export license	Own processing and marketing, has export license
Other services	None	Collection stations & marketing of grains, legal support	Fertilizer production, transports, wet processing of coffee	Legal assistance available via 2nd level cooperative	None	None
Coffee extension per year	Monthly group trainings, (exporter)	Monthly group trainings	Monthly group trainings	4 group trainings and 3-6 farm visits	6 group trainings & monthly farm visits	20 group trainings & monthly farm visits
Extensionist / producer ratio	1 / 113	1 / 248	1 / 11	1 / 120	1 / 28	1 / 28
Financing of extension	Gov't. institute	NGO	Self-financed & NGO	NGO	NGO, gov't. institute	NGO, gov't. institute
Coffee contracts with members	No, only via credit	No	No, only via credit	No, only via credit	No, only via credit	No, only via credit
Compliance of members with credit contracts	No; 60% default as bad coffee year	Yes, 95%	Yes	Default down from 34% to 8% (2008)	Usually yes (90%)	Yes, around 80%.
Cooperative growth strategy	Stability in members, production & sales	Slower growth of members; stability in production & sales	Growth of coop., coffee production & sales	Growth of coffee production & sales	Growth of coffee production & sales	Growth of coop, coffee production & sales
Major income of coop.	Through loan interests	Through loan interests	Coffee sales	Coffee sales	Coffee sales	Coffee sales
Profits available	No	Yes	Yes but small	Yes	Yes	Yes
Reserve fund	Yes	Yes	No	Yes, small	Yes, very small	Yes
Credits obtained from...	Exporter	Intl. (ethical) banks	Exporter	2nd-level cooperative	Exporter	Intl. (ethical) banks
Sufficient credit funds	No	Yes	No	Lack 500,000US$	(Yes) Insufficient for member credits	Lack around 250,000US$

inputs, storage, and grain as well as coffee marketing. The certified cooperatives focus mainly on service provision regarding coffee-related services like processing, marketing, and extension. Marketing is done by the cooperatives themselves or by the second-level cooperatives with which they have alliances. Except for the conventional cooperative CONV-A, members could obtain basic farm inputs from their cooperatives. The organic cooperative ORG-C additionally produces and sells a high quality organic fertilizer at cost-covering prices. For a small fee the cooperative even transports the fertilizer to the member's plots. Through both services members are able to apply more organic fertilizer than normal which helps to increase their yield levels (see Table 6.6). The same cooperative also built a central wet processing station[2] for their members' coffee in 2006. This was an essential upgrading strategy as many members did not individually own wet processing mills.

The extension service for coffee producers varies significantly between cooperatives. The worst extensionist-producer ratio was offered by the cooperative CONV-B / ORG-B for both their conventional and organic members (Table 6.3). The best extensionist-producer ratio is offered by the two smaller Organic-Fairtrade certified cooperatives which received funding from both an NGO and a governmental institution for the extension service. Extension relates mainly to coffee production issues. Field observations and qualitative interviews showed that the need for enhancing farm management and business skills of smallholder producers is very high. Yet training on this is hardly done because the extension funding organizations do not cover it in their guidelines. As cooperative ORG-C could not obtain sufficient external funding for an extensionist, the members decided to pay for it themselves.

Coffee producers do not have contracts regarding coffee delivery or prices with their cooperative. Only when taking a credit from their cooperative are members obliged to pay the credit with coffee, except in cooperative CONV-B. Through this tied contract the cooperatives ensure loan repayment and delivery. They can also estimate the minimum coffee quantity delivered to the cooperative that will reduce the risk when negotiating with coffee buyers. Apart from the conventional cooperatives, the cooperative credit supply for their members is limited by a shortage of funds and liquidity constraints of the certified cooperatives.

The certified cooperatives focus on growth of coffee production and thus coffee sales with the aim of reducing transaction costs through economies of scale. The two conventional cooperatives focus on stability in production and sales. Cooperative CONV-B experienced strong member growth rates in the past five years; in 2003, they had only around 1,400 members compared with 9,300 in 2008. Apart from having opened many regional branch offices which contributed to their growth, the cooperative also introduced an insurance policy to cover funeral costs for their members, which, according to informal statements, attracted many new members.

---

[2] In coffee production, a wet processing station or mill processes the coffee cherries by removing the coffee pulp from the coffee beans. Afterwards, the coffee is washed and dried.

The cooperatives are not financially autonomous. Apart from cooperative CONV-B, all depend on credit and face limited credit access. Cooperatives often lack appropriate collateral and therefore, they do not obtain sufficient credits to satisfactorily manage the coffee marketing, investments or the members' credit demand. Cooperatives have no access to state bank credit which could help to reduce their dependency on exporters and international finance institutions. Cooperatives CONV-B/ORG-B and OFT-F obtain credit from international ethically oriented finance institutions such as Oikocredit or Rabobank. Since 2004 cooperatives are required by law to have a reserve fund to increase their financial stability but cooperatives struggle to build up this fund. Usually credit repayment rates of members are relatively good, but in 2007/08 unfavorable microclimatic conditions caused a small coffee harvest. Cooperative CONV-A was especially hard hit by high credit default rates of its members.

Cooperatives apply several upgrading strategies in order to add more value to their members' coffee and to improve the generated incomes of the cooperative and its members. Table 6.4 categorizes the different upgrading strategies of the cooperatives based on the conceptual framework of Riisgaard et al. (2010) (cf. Section 6.2). Cooperatives concentrate on product, volume, and functional upgrade as well as vertical coordination. They often use the same upgrading strategies. Only two cooperatives try to differentiate from their competitors. The cooperative ORG-C offers most in terms of process and functional upgrade for its members through offering fertilizer and wet mill services. The cooperative OFT-F improved its coffee quality and services for its coffee buyers through a side office and increased traceability. Apart from coffee, the funeral insurance of cooperative CONV-B can be seen as another upgrading strategy which successfully attracted many new members. The functional upgrading to ecotourism of cooperative OFT-D is an imitation of another coffee cooperative in another department. In both areas, it was not successful and resulted in a very costly failure for OFT-D. No cooperative has considered downgrading as an option so far. While there is a tendency of cooperatives to become involved in more and more upgrading activities, the additional benefit of another upgrading strategy is often not properly evaluated by cost-benefit analyses. The many upgrading strategies of cooperative OFT-D became a large financial and management burden for the cooperative. A strategy of downgrading could reduce this burden, for example, through the cooperative dropping the ecotourism activities which are costly and without benefits. It is also questionable whether roasting, the registration of its own coffee brand and local marketing is worthwhile as it consumes many human and financial resources, a clear marketing strategy for its coffee is lacking and competition in this area is high. Supported by NGOs which finance part of the roasting equipment, many cooperatives now offer roasted coffee despite the facts that the non-coffee growing national population prefers instant coffee and the export market unroasted coffee.

The planned change in horizontal value chain coordination of cooperative OFT-F to reduce export related transport costs through a transport alliance with other cooperatives

Table 6.4.: Upgrading strategies of conventional, organic and Organic-Fairtrade certified cooperative models

Cooperative	CONV-A	CONV-B / ORG-B	ORG-C	OFT-D	OFT-E	OFT-F
			Product upgrade			
(1) Improvement of the product, production process, or volume	• *Certification is planned* • *Quality project is planned* • *Improved production, harvest*	• Organic production • Distinct quality profiles • *Further certifications are planned*	• Organic production • High quality • Improved production, harvest, processing	• Organic production • Fairtrade certification • Distinct quality profiles • Improved production, harvest, processing	• Organic production • Fairtrade certification • High quality • Improved production, harvest, processing	• Organic production • Fairtrade certification • Distinct and high quality profiles • Improved production, harvest, processing
			Process upgrade			
	• N.a.	• N.a.	• Improved production efficiency as own fertilizer production • Organic waste recycling into fertilizer	• N.a.	• Improved invoicing for producers	• Improved services for coffee buyers through side office • Traceability of coffee
			Volume upgrade			
	• No	• Yes	• No	• Yes	• Yes	• Yes
			Functional upgrade			
(2) Change and/or the adding of functions	• Commercial credits	• Input supply • Own dry mill • Own roasting & own local brand • Commercial credits • Agric. diversification • Funeral insurance	• Input supply • Own wet mill • Own roasting & local brand (youth activity)	• Input supply • Cupping laboratory • Own dry mill • Own roasting & own local brand • (limited) agric. diversification • Ecotourism	• Input supply • Own local brand	• Input supply • Cupping laboratory • Own dry mill
			Vertical coordination			
(3) Value chain coordination	• Only with exporter	• Yes. Improved marketing through 2nd-level coop.	• Yes, starting. Improved marketing through alliances	• Marketing with 2nd-level coop, direct contacts to roaster	• Yes, direct contacts to coffee importers/roasters	• Yes, direct contacts to coffee importers/roasters
			Horizontal coordination			
	• N.a.	• N.a.	• N.a.	• N.a.	• N.a.	• *Planned transport alliance with other coops. to reduce costs*

is a opportunity if it is well managed and regulated. In general, most upgrading strategies can be easily imitated by other cooperatives. Consequently, successful models of one cooperative are copied by other cooperatives and recommended by policy-makers and NGOs. As cooperatives have different business models, the imitating cooperative is not always likely to be successful and additionally creates competition in what are often very small market niches. Due to the ease of imitation, returns to investments of cooperatives and their members are reduced to only a short-term advantage. Additionally, it enables the coffee buyers to exert more bargaining power and to put pressure on prices, for example when many cooperatives offer certified coffee (cf. conceptual framework). This is already felt by the certified cooperatives which face declining organic premiums but rising quality demands from their buyers.

6.5. Strengths, weaknesses, opportunities, and threats of cooperatives

There are strengths, weaknesses, opportunities, and threats (SWOTs) which are common to all cooperatives and SWOTs which are specific to individual cooperatives. The SWOTs of each of the conventional and certified cooperatives are illustrated in Table 6.5. Elements which are a strength of one cooperative can be a weakness or threat to another cooperative. For example, this is the case for the internal organization, staff commitment, or the extension system. We discuss in detail the SWOTs common to all cooperatives and derive a number of implications for management changes at the cooperative level and other value chain actors (buyers, banks, NGOs) as well as for policy. The specific needs of a cooperative should be addressed when donors and NGOs are involved, as policy changes alone cannot reduce the cooperative's specific weaknesses and threats. The **strength** of all cooperatives is that their coffee growing members are located in high altitudes with a very good coffee quality potential. The region's coffee has won several prizes in international recognized and highly prestigious competitions. Generally, all cooperatives indicate a trend in the coffee industry towards higher quality, especially those having certification. Otherwise, they do not expect major changes in the coffee industry or national governmental regulations in the near future. There is a relative **stability for planning** after taking into account the coffee price volatility. Investing in high quality constitutes an **opportunity** for coffee growers and cooperatives. Policies could be directed at strengthening the coffee quality in this region and among cooperatives, and advertising the quality internationally; they have already started by establishing an internationally famous cupping event of high quality coffee in Nicaragua (The Cup of Excellence). While cooperatives had good results in the beginning, in recent years they lost the top positions and their participation decreased.

There are several **weaknesses** common to all cooperatives. First, they strongly depend on external **credit supply**. Nearly all cooperatives **lack own funds or credit**

Table 6.5.: SWOT of conventional, organic and Organic-Fairtrade certified cooperative business models

	Strengths	Weaknesses	Opportunities	Threats
CONV-A	• Mainly cost-recovering interest rates • Good treatment of members • Good relationship with exporter • Committed manager	• No business plan • Dependence on exporter • Insufficient coffee commercialization service External credit supply insufficient • Uncommitted field staff	• Improving coffee quality • Increase extension • Find committed field staff • Improve leadership of manager • Find a coffee market alliances	• Strong competition from other coops/intermediaries buying coffee ⇒ contract breach with exporter
CONV-B / ORG-B	• Many services, incl. radio channel & shop • Mainly cost-recovering interest rates • Good coordination betw. administration / field staff • Good organizational model, good commitment • Good organic coffee quality • Internet access at all times • Good credit access ⇒ intl. ethical banks	• Strong growth in members ⇒ members loose identification with coop. • Support for organic producers very low ⇒ dissatisfaction with coop. and farming method • Insufficient extensionists and support for coffee producers	• Improving coffee marketing through 2nd level coop. to reduce transaction costs • Built up of strong extension system for their agricultural producers • Restructure towards more small sized groups with regular meetings	• Not up-keeping with their services due to fast growth • Neglect of other services to members except finance ⇒ competition from other microfinance institutions
ORG-C	• Strong leadership • Strong support of members in case of problems • Own fertilizer production & transport to farmers • Strong cooperative values of leaders and members ⇒ high commitment & good treatment • High coffee quality • Alliance with exporter who offers good prices • Only weak competition from other coffee buyers • Management done by members themselves	• No skilled financing / marketing person • Limited marketing knowledge • No business plan • Administration by hand • Bad infrastructure (no phone, internet, electricity...) • No own vehicle or truck • Credit supply insufficient to finance producers • Small average land size	• Get high quality better rewarded through additional market alliances • Coffee quality tests to increase bargaining possibility with exporter • Invest in higher education of representative coop members to support financing/marketing • Open a mini office in a nearby village with infrastructure	• Coffee price volatility • Strong dependence on exporter • Land constraints and high poverty among members ⇒ neglect coffee fields in search for additional income opportunities • Problems with obtaining land title for the cooperative's members

to be continued ...

Table 6.5.: SWOT of conventional, organic and Organic-Fairtrade certified cooperative business models (–continued–)

	Strengths	Weaknesses	Opportunities	Threats
OFT-D	• Coffee sorted according to quality • Good support by buyers • Good "self-marketing" skills • Internet access in office • Own coffee brand • Offer different production technologies	• Unmotivated extensionists • Financial mismanagement ⇒ debts • Members unhappy with organic farming • Low coffee prices paid to producers • Internal organizational model problematic • Oversized investments	• Better extension • Improve payment schemes to producers • Increase transparency • Downgrade in functions • Restructure internal organization in smaller groups	• Dependence on donors for extension • Producers quit and/or increase off-sales ⇒ contract noncompliance • Loss of certification • Continuing irregularities in cooperative ⇒ further indebtedness
OFT-E	• Good prices for producers • Issue premiums • Good quality • High transparency in producer bills • Basic organic input provision • One very committed head of extensionists	• Difficulties for buyers to contact coop ⇒ slow responding of e-mails • Internal staff problems • Insufficient credit possibilities for producers • High processing & export costs • Cannot offer good spot market prices	• Ally with other cooperatives to lower transport and marketing costs • Change subcontracted dry mill • Stronger internal management / leadership	• Dependence of external financing of extension • High world market prices ⇒ off-sales ⇒ no contract compliance • Loss of committed staff
OFT-F	• Very committed, producers and cooperative staff • Transparency in their activities • High quality of coffee ⇒ strong quality controls • 2nd office in other city ⇒ good contact with buyers • Coop. staff are often members themselves • Support of producers in case of emergency • Strong efforts to increase producer revenues • Credit supply from international ethical banks	• Coffee commercialization costs are high, especially transport costs to harbor • Low yields despite many efforts from cooperative • Credit supply insufficient as collateral lacks • Demand for high quality raises costs for producers ⇒ off-sales • Low average land size of members	• Ally with other cooperatives regarding transport to lower commercialization costs • Extend cupping laboratory • Enter more the gourmet quality market • Stronger focus on replanting/improving coffee plantations	• Dependence on external financing of extension • High world market prices ⇒ off-sales ⇒ no contract compliance ⇒ loss of buyers

access to finance all their operations and to offer adequate credit to their members. As cooperatives cannot provide enough credit, their members obtain additional credit from other sources like microfinance institutions, of which a few operate in the research region, or from intermediaries which offer a cash advance on coffee delivered later. The coffee delivered to intermediaries is then not available to the cooperative which reduces its turnover. Policies could address the liquidity constraints of cooperatives through, for example, a supportive national banking system or an agricultural bank which provides business credits at market interest rates to the cooperatives. Additionally, policies could promote a diversified microfinance sector to serving the cooperatives' members. The certified cooperatives should rethink the credit provision to their members; it creates a high administrative burden and requires human capacities. The whole cooperative faces the risk of credit defaults through their members (see Table 6.3) and credit for coffee carries excessive covariate risk with respect to yields and climate, prices, and competition. Such risk can be better shouldered by international finance institutions or by broadly diversified, national microfinance and banking institutions (Zeller et al., 1997). Cooperatives could, for example, collaborate with microfinance institutions which provide the credit to the cooperative's members while the cooperative sells the member's coffee, then first pays the member's credit at the microfinance institution before paying the remaining gains from coffee sales to the credit using member. On the other hand, microfinance institutions could serve cooperatives at relatively low transaction costs as the cooperatives themselves could undertake the screening, monitoring and loan enforcement functions. For banking and micro-finance institutions, working with rural cooperatives could thus enhance their business exposure to rural areas, lead to risk-reducing diversification of their lending portfolio, and enhance their social performance by providing credit to low-income male and female coffee farmers. Cooperatives must improve their liquidity position and credit worthiness over time by building up more equity capital and their reserve fund. This process should be made transparent so that the members are aware that their shares in their cooperative are increasing.

The cooperatives are located in an area with a **weak rural infrastructure.** This raises the cooperative's operation costs, e.g., expenses for an electrical generator, transport problems due to unpaved roads in bad condition, unreliable fixed and poor mobile phone and internet connectivity requiring trips to other towns to contact buyers. Policies in this field should target investments in roads and electricity. Reliable phone and internet connections are important to enable the international marketing of the coffee through the cooperatives.

None of the cooperatives have completely self-financed extension service; for this too they depend strongly on external funding. As **extension** is financed by short-term development projects, within a couple of years it is probable that extension services will no longer be offered, negatively affecting the producer's commitment. It is recommended that cooperatives follow the model of ORG-C in which producers participate in financing extension.

A **threat** for cooperatives is the **microclimatic variation** in the research area, which, according to the interviews, is increasing. Coffee producers indicated that weather conditions are becoming less predictable. Coffee, but especially the coffee flower, requires a certain rainfall pattern. With increasing climate variability due to climate change, rainfall or drought at the wrong phenological stage can lead to severe harvest losses. The microclimatic variations make it difficult for cooperatives to estimate the harvest correctly. Wrong estimates can threaten the compliance with coffee and credit contracts and raise transaction costs in the pursuit of missing coffee. It is difficult for farmers to deal with the unreliable microclimate; they can reduce their risks by diversifying into other agricultural crops, improving shade management in the coffee plantation and opting for less climate susceptible coffee varieties. The government could support research activities for high-yielding coffee varieties which are less susceptible to unfavorable climatic conditions.

Another **threat** to the cooperatives is the intense **competition** from other coffee buyers like intermediaries, exporters and other cooperatives - the organic cooperative ORG-C being the only exception. The advantage of the non-cooperative buyers in Nicaragua usually is that they possess sufficient funds to pay producers on the spot, which is difficult for cooperatives that face liquidity constraints at harvest time. Additionally, exporters can easily fix coffee on the stock exchange when prices are high and offer these prices to producers. When members are dissatisfied with spot market prices, they sell part of their coffee elsewhere. Off-sales of coffee can threaten cooperatives when producers start to default on the cooperative credit and contracted coffee quantity. Short-term bank credits, from a national bank, could help to ease the liquidity constraints of cooperatives. Microfinance institutions could provide short-term consumption credits to producers during harvest time to reduce their need for spot market sales.

Certified cooperatives are further threatened by **high conventional coffee prices**. In times of increasing world market prices, the additional premiums for certified coffees dwindle as conventional coffee prices rise at a faster rate than organic and Fairtrade coffee prices. The buyer expectations of higher quality standards for certified coffee remain the same. Since high quality coffee is costly to produce, and the returns are diminishing, some cooperative members are choosing to produce lower quality coffee and increase their off-sales. As the CEO of cooperative OFT-F put it: *"It sounds like a paradox, but the biggest threat for us is high world market coffee prices"* (personal communication, April 2008).

Another threat to all cooperatives is the possibility of **corruption and mismanagement** of funds. Because boards or executive managers run the cooperatives' businesses and make important decisions, members of cooperatives have little or no control and are usually only informed of changes at the annual assembly. Nearly all the cooperatives interviewed had severe management and/or corruption problems in the past 10 years, other non-interviewed cooperatives even went bankrupt. Liability regulations in the cooperatives make members responsible for paying the debt. The introduction of an obligatory annual external auditing of first-level and second-level cooperatives is recommended to avoid mismanagement and corruption and to make cooperatives more creditworthy for banks. The external auditors should be subjected to further external quality controls, as well as be controlled by a governmental supervisory authority.

As members often are not able to see all benefits of their cooperative, one possible solution is to develop and implement the concept of a **member's promotion plan**, designed by the cooperative's management and board together with the members. This is an **opportunity** for all cooperatives, and ORG-C seems to already be utilizing such a strategy well, although not in written form. All goals of this promotion plan should be verifiable; the plan should be provided at the beginning of the business year and evaluated at the annual assembly. Additionally, there could be an annual record of each member's cumulative capital contributions (e.g., the association fee or the percent retained from credits for reserve funds) and some sort of **share certificate acknowledging the members' contribution**. While this involves human and financial resources, these tangible documents enhance transparency and could increase members' loyalty and satisfaction with the cooperative more than any oral statements on the benefits of cooperative membership.

Another opportunity is to increase the **transparency of the deductions** on the coffee which members receive as final bill for their delivered coffee. Although cooperatives briefly present that information in the annual assembly, it should be clearly visible on the invoice, expressed in simple words, listing detailed information of the total payment per unit coffee made by the coffee buyer and the deductions for processing, export, investments, and certification. Credit repayments should be issued on a separate invoice.

In order to reduce transaction costs, a greater horizontal coordination among cooperatives, like the planned transport alliance of the cooperative OFT-E, is another opportunity for the cooperatives. Alliances do not only need to be among cooperatives, but could also be with exporters, through, for example, the hiring of dry mill services.

Until 2008, government support of cooperatives, agricultural producers, and coffee smallholders was negligible. Producers from other countries like Colombia, Costa Rica, Brazil or Mexico can count on governmental strategy and support, not only through credit, extension, subsidies, market support or infrastructure, but also through research and general higher educational levels. The absence of this support and the lack of a functioning national coffee institution, like the federation of coffee growers in Colombia (FNC) or the coffee institute in Costa Rica (ICAFE), make it difficult for Nicaragua's cooperatives and their members to place themselves in the international coffee sector and maintain their competitiveness. They need additional support from a national organization with a long-term focus. The coffee federation or institute could also be responsible for the monitoring of finances or the auditing of cooperatives and could forge links with the national banking system and international banks.

6.6. Linking gross margins to business models, certification, and upgrading strategies

The average yield levels and gross margins of coffee producers are shown in Table 6.6. The members of the two conventional cooperatives on average received a similar coffee price across all market channels. Cooperative CONV-B paid its members much higher coffee prices than the exporter of CONV-A. The yield level is higher in cooperative CONV-B, and production costs per hectare are similar between the two conventional cooperatives, resulting in 90US\$/ha higher gross margins for producers in CONV-B compared with CONV-A.

Between the two organic certified cooperatives there is a high yield variation. The yield per hectare of cooperative ORG-B is nearly half compared with ORG-C. The average coffee prices producers obtained across all market channels are similar between the two cooperatives but cooperative ORG-C paid significantly higher coffee prices to its members. Although producers of ORG-C have higher variable production costs, their gross margins are still double that of ORG-B. Some reasons behind the significantly higher yield of cooperative ORG-C's coffee producers are that they better manage their coffee plantations and they have good access to organic fertilizer. The members of the Organic-Fairtrade certified cooperatives have lower yields than the organic cooperative ORG-C. Even among them, there are differences in yield levels, coffee prices and production costs. Members of the cooperative OFT-D have nearly the same average coffee price as producers in the conventional cooperative CONV-B, while the other Organic-Fairtrade producers obtain better average prices. The highest prices paid by the cooperatives are similar among all three Organic-Fairtrade cooperatives. Members of cooperative OFT-D either had higher off-sales than the other cooperative members or received more spot market prices with slight premiums by their cooperative instead of being issued a final bill. Yield levels and production costs are relatively similar between cooperative OFT-D and OFT-F while members of cooperative OFT-E have higher yields and higher production costs. Among the Organic-Fairtrade cooperatives, members of OFT-E have the highest gross margins, earning 62% more than members of cooperative OFT-D. Members of OFT-D obtained the lowest per hectare gross margins of all the cooperatives. Its members also showed dissatisfaction with the organic farming system and their cooperative during the interviews. The low coffee prices and low gross margins in OFT-D reflect the above described management problems, failing upgrading strategies and weaknesses of the cooperative. While members in the other Organic-Fairtrade cooperative, OFT-F, received only 40US\$/ha more, they were satisfied with the prices, their cooperative and its services. They mentioned low yields as their one major problems caused by old and low-density planted plots. They aim to improve their plots but lack access to credits for long-term investments like the replanting of coffee plots.

Table 6.6.: Coffee prices (US$/kg) and gross margins per hectare coffee of conventional, organic and Organic-Fairtrade certified cooperatives

Cooperative	CONV-A	CONV -B	ORG-B	ORG -C	OFT-D	OFT-E	OFT-F
Yield of green coffee (kg/ha)	333.0 ± 180.7	393.6 ± 200.1	293.1 ± 223.7 [b]	515.8 ± 235.0 [b]	292.3 ± 106.5	411.2 ± 200.9	306.2 ± 148.3
Average farm-gate coffee price (US$/kg)	2.08 ± 0.20	2.12 ± 0.27	2.29 ± 0.15	2.28 ± 0.20	2.14 ± 0.46	2.38 ± 0.16	2.35 ± 0.26
Highest farm-gate coffee price paid by the members' cooperative (US$/kg)	2.06 ± 0.19[b]	2.19 ± 0.13[b]	2.31 ± 0.13[b]	2.37 ± 0.11[b]	2.39 ± 0.12	2.42 ± 0.16	2.47 ± 0.23
Revenue (US$/ha)	**693.2 ± 372.8**	**821.3 ± 409.5**	**666.9 ± 503.7** [b]	**1172.5 ± 555.7** [b]	**639.4 ± 297.2** [b]	**975.5 ± 464.6** [ab]	**711.9 ± 338.9** [a]
Variable production costs (US$/ha)	283.9 ± 180.6	275.4 ± 139.4	205.1 ± 130.1	268.3 ± 231.1	251.3 ± 203.6	346.6 ± 272.1	283.4 ± 212.1
Gross margin (US$/ha)	**454.2 ± 320.9**	**545.9 ± 371.5**	**461.8 ± 437.6** [b]	**904.2 ± 487.5** [b]	**388.1 ± 278.2** [b]	**628.9 ± 346.4** [b]	**428.5 ± 276.0**

Note: [a] Significant difference at p<0.05. [b] Significant difference at p<0.01. Superscript letters indicate a significant difference between two groups marked by the same letter. Prices and gross margins statistically compared within a certification status, i.e., between CONV-A and CONV-B, between ORG-B and ORG-C, and between OFT-D, OFT-E, and OFT-F. All variables were non-normally distributed. Kruskal-Wallis tests were used to test for statistical differences, followed by pairwise comparisons based on the Mann-Whitney post hoc test adjusted by the Bonferroni correction factor where applicable (Field, 2005).

6.6. Linking gross margins to business models, certification, and upgrading strategies

The highest gross margin is obtained by the organic cooperative ORG-C followed by the Organic-Fairtrade certified cooperative OFT-E. Paradoxically, the two lowest gross margins are also obtained by Organic-Fairtrade certified cooperatives (OFT-D, OFT-F). Members of conventional cooperatives have gross margins somewhere between those of the certified cooperatives. Table 6.7 ranks the cooperatives according to the gross margins earned by the cooperative members, by elements of the cooperative's business models, and upgrading strategies. There is no obvious association between the certification strategy of

Table 6.7.: Relative ranking of conventional, organic and Organic-Fairtrade certified cooperatives

	CONV-A	CONV-B / ORG-B	ORG-C	OFT-D	OFT-E	OFT-F
Average gross margins	Middle	Middle	High	Low	High	Low
Upgrading						
Certification as upgrading strategy	n.a.	n.a. / Organic	Organic	Organic-Fairtrade	Organic-Fairtrade	Organic-Fairtrade
No. of other upgrading strategies	Low	High	Middle	High	High	High
a) Coffee-related services offered	Very few	Few	Many	Many	Many	Many
b) Quality of these services	Basic	Medium	Very good	Low	Very good	Very good
Business model						
Size of cooperative	Middle	Large/ Small	Small	Large	Middle	Small
Financial situation & characteristics of cooperative	Middle	Very good	Weak	Weak	Good	Good
No. & importance of strengths	Few	Middle	Many	Few	Middle-Many	Many
No. & importance of weaknesses	Middle	Middle/ Many	Middle	Many	Few	Few
Members' satisfaction & commitment to coop[*].	Low	Low	High	Low	Middle	High

[*] Only indirectly measured, based on qualitative interviews, informal comments and observations. ORG-C and OFT-F offer many 'informal services' such as helping members incapable of paying their debts, providing transport to hospital, granting collateral free credit in case of severe illness/accident etc.

a cooperative and the gross margins. Producers in certified cooperatives may have either higher or lower gross margins than producers in conventional cooperatives. Members' satisfaction as well as the net of strengths and weaknesses of a cooperative can be related to the gross margin. The data indicate that OFT-F is an exception, as it is a very well-run and organized cooperative but has very low gross margins for members. The example of the cooperative ORG-C shows that its many weaknesses can be compensated by its strengths and coffee-related services, resulting in high member satisfaction and contributing to good gross margins. The number and importance of strengths and weaknesses as well as the amount of coffee-related services offered to producers seem to affect the gross margins more than certification. Table 6.7 also shows no clear relationship between the certification strategy of a cooperative and the number of other upgrading strategies, cooperative size, and financial characteristics. This qualitative evaluation indicates that the gross margins per hectare depend on the business model of a cooperative, its respective SWOTs, and all upgrading strategies together. The organic or fairtrade certification alone as an upgrading strategy is not a sufficient for obtaining higher coffee gross margins.

6.7. Conclusion

In this paper, we first analyzed the business models and upgrading strategies of coffee cooperatives in northern Nicaragua. Each cooperative has a unique business model; they differ, for example, in member size, functions and services, internal organization, and financial characteristics. Despite their different business models the cooperatives often choose the same upgrading strategies as other cooperatives, choosing, for example, to upgrade their product coffee through certification and high quality, or to upgrade their functions through running a dry mill, branding their coffee, or ecotourism. When these upgrading strategies do not fit the business model of a cooperative, especially its financial and human/management capacities, they can negatively affect the performance of the cooperative. The growing popularity of environmental and social certifications among cooperatives, like organic and Fairtrade certification, as well as the imitation of other upgrading strategies are leading to increased competition with negative effects among Nicaragua's cooperatives. Cooperatives, therefore, need to identify specific upgrading strategies and market niches which do not overburden the cooperative's management or finances. Downgrading of activities without clear benefits, like ecotourism, can also be a valuable strategy for some cooperatives.

Second, we analyzed the cooperatives' strengths, weaknesses, opportunities, and threats as these are related to the cooperative's business model. We identified SWOTs common to all cooperatives and also cooperative specific SWOTs. The latter need to be addressed internally by the management of the cooperative and can perhaps be supported by vertical coordination and other chain actors like banks or coffee buyers. Common SWOTs can be addressed by policy. The common strength of the cooperatives is the quality potential in the region. The common weaknesses are many and relate to the lack of sufficient credit access of cooperatives, a weak extension system, and weak rural infrastructure. The common threats of the cooperatives are high competition among national coffee buyers and cooperatives, corruption and mismanagement, increasing microclimatic variations and climate change. For certified cooperatives, high conventional coffee prices pose another threat as the premiums paid for certified coffee do not rise proportionally. The common opportunities range from horizontal coordination for further reduction of transaction costs to enhancing members' satisfaction and commitment to their cooperative through a member's promotion plan, transparency about deductions on payments and share certificates acknowledging the member's contribution and possessions.

The third part of this article addressed the coffee gross margins and how they are linked to the cooperative's business model, its product upgrade through organic and Fairtrade certification and its other upgrading strategies. We found no obvious association between the coffee certification strategy of a cooperative and the coffee gross margins of its members. The number and importance of strengths and weaknesses as well as the amount of coffee-related services offered to producers tend to be more related to gross margins than the organic or Organic-Fairtrade certification. Based on a qualitative evaluation, we conclude that the gross margins per hectare depend on the business model of a cooperative and thus the services it offers, its respective SWOTs, and upgrading strategies of which the certification strategy is one. Organic or Organic-Fairtrade certification as an upgrading strategy seems only then successful when the business model of a cooperative, its strengths, weaknesses, and other upgrading strategies are supportive. More research on this topic is necessary and future analysis should be based on a large quantitative data set allowing for statistical and causal analyses. Impact assessments of certification schemes on the socio-economic effects on smallholders should include the institutional context in which farmers operate through explicitly analyzing the cooperative's business models and its upgrading strategies. Comparisons across countries which include the institutional setting at national level will be especially valuable.

It is recommended that policies which aim at increasing smallholder coffee incomes through upgrading should focus, apart from production aspects, on the institutional context of smallholders and their cooperatives. Regarding coffee production, policies could be directed at increasing yields, for example through improved and stress-tolerant varieties, and strengthening the coffee quality in the region. While not having been discussed in detail, a closer look at successful policies in the coffee sector from other countries, like Colombia's coffee federation, reveals that institutional support at national level is of great importance. Competitive advantage of smallholders and their cooperatives is not built over night, but must be based on long-term strategic development of the whole national coffee sector. With limited demand in consuming countries, coffee certifications alone cannot help to improve producers' welfare in the long run because entry barriers are low and new producers will continue to enter until the profit margin is zero. Therefore, a supportive sector strategy, focusing also on quality, needs to be developed and effectively implemented at the national level. This requires investments in rural infrastructure, such as electricity grids, roads, internet and phone connections. Coffee producers should have access to extension services which are independent of short-term development projects and which focus on agricultural diversification and business skills of coffee producers. This could be in the form of an extension association, operating regionally, which is financed by its members' contribution.

Business and strategic advice to cooperatives is necessary, as often the education level of cooperative leaders and staff is not internationally comparable, yet they have to act in international markets. A national or agricultural bank which provides credits to co-operatives (at market interest rates and lending conditions) would reduce the reliance and dependence on exporters or international credit providers and could contribute to loosening liquidity constraints of cooperatives while diversifying the loan portfolio and enhancing the social performance of the bank. We consider very important the introduction of an obligatory annual external auditing of cooperatives to avoid mismanagement. This will also increase the creditworthiness of cooperatives for banks.

Policies could further oblige cooperatives to be more transparent in their invoices and to issue a record of each member's cumulative capital contributions or share certificates acknowledging the member's possession in the cooperative. This enhances transparency and members' control of their cooperative but also benefits the cooperative as members become more satisfied and committed. These measures could be implemented voluntarily by cooperatives to reduce coffee off-sales by their members and thus improve planning and marketing, reducing the risk that cooperatives may not be able to comply with contracts. The members received more security for their cooperative investments and can better compare the benefits of their cooperative with others. Cooperative members thus reduce their information asymmetries, allowing them to make more informed decisions about membership, upgrading of production systems, certification, and marketing.

6.8. Bibliography

Arnould, E. J., Plastina, A., Ball, D., 2009. Does fair trade deliver on its core value proposition? Effects on income, educational attainment, and health in three countries. Journal of Public Policy & Marketing 28 (2), 186–201.

Asfaw, S., Mithöfer, D., Waibel, H., 2010. Agrifood supply chain, private-sector standards, and farmers' health: Evidence from Kenya. Agricultural Economics 41 (3-4), 251–263, 1574-0862.

Bacon, C., 2005. Confronting the coffee crisis: Can fair trade, organic and specialty coffees reduce small-scale farmer vulnerability in northern Nicaragua? World Development 33 (3), 497–511.

Bacon, C., Ernesto Mendez, V., Gomez, M. E. F., Stuart, D., Flores, S. R. D., 2008. Are sustainable coffee certifications enough to secure farmer livelihoods? The Millennium Development Goals and Nicaragua's fair trade cooperatives. Globalizations 5 (2), 259–274.

Barham, B. L., Callenes, M., Gitter, S., Lewis, J., Weber, J., 2011. Fair trade/organic coffee, rural livelihoods, and the "agrarian question": Southern Mexican coffee families in transition. World Development 39 (1), 134–145.

Barrett, H. R., Browne, A. W., Harris, P. J. C., Cadoret, K., 2001. Smallholder farmers and organic certification: Accessing the EU market from the developing world. Biological Agriculture and Horticulture 19 (2), 183–199.

Basu, A. K., Chau, N. H., Grote, U., 2003. Eco-labeling and stages of development. Review of Development Economics 7 (2), 228–247.

Bolwig, S., Gibbon, P., Jones, S., 2009. The economics of smallholder organic contract farming in tropical Africa. World Development 37 (6), 1094–1104.

Bolwig, S., Ponte, S., Du Toit, A., Riisgaard, L., Halberg, N., 2010. Integrating poverty and environmental concerns into value-chain analysis: A conceptual framework. Development Policy Review 28 (2), 173–194.

Bray, D. B., Sánchez, J. L. P., Murphy, E. C., 2002. Social dimensions of organic coffee production in Mexico: Lessons for eco-labeling initiatives. Society & Natural Resources 15 (5), 429–446.

Cashin, P., McDermott, C. J., Scott, A., 2002. Booms and slumps in world commodity prices. Journal of Development Economics 69 (1), 277–296.

Codron, J.-M., Siriex, L., Reardon, T., 2006. Social and environmental attributes of food products in an emerging mass market: Challenges of signaling and consumer perception, with European illustrations. Agriculture and Human Values 23 (3), 283–297.

Dabbert, S., Braun, J., 2009. Landwirtschaftliche Betriebslehre, 2nd Edition. Eugen Ulmer, Stuttgart.

Deininger, K., 1995. Collective agricultural production: A solution for transition economies? World Development 23 (8), 1317–1334.

Dülfer, E., 1995. Betriebswirtschaftslehre der Genossenschaften und vergleichbarer Kooperative, 2nd Edition. Vandenhoeck und Ruprecht, Göttingen.

Field, A., 2005. Discovering statistics using SPSS, 2nd Edition. SAGE Publications, London, Thousand Oaks, New Delhi.

Fleisher, C. S., Bensoussan, B. E., 2003. Strategic and competitive analysis: Methods and techniques for analyzing business competition. Prentice Hall, Upper Saddle River, US.

Gereffi, G., 1999. International trade and industrial upgrading in the apparel commodity chain. Journal of International Economics 48 (1), 37–70.

Giuliani, E., Pietrobelli, C., Rabellotti, R., 2005. Upgrading in global value chains: Lessons from Latin American clusters. World Development 33 (4), 549–573.

Hellin, J., Lundy, M., Meijer, M., 2009. Farmer organization, collective action and market access in Meso-America. Food Policy 34 (1), 16–22.

Humphrey, J., Schmitz, H., 2002. Developing country firms in the world economy: Governance and upgrading in global value chains. INEF Report 61. Institut für Entwicklung und Frieden der Gerhard-Mercator-Universität Duisburg, Duisburg.

IFOAM, 2006a. Organic agriculture and food security. International Federation of Organic Agriculture Movements (IFOAM), Bonn, accessed 20 July 2006.
URL `http://www.ifoam.org/organic_facts/food/pdfs/Food_Security_Leaflet.pdf`.

IFOAM, 2006b. Organic agriculture and rural development. International Federation of Organic Agriculture Movements (IFOAM), Bonn, accessed 15 December 2010.
URL `http://www.ifoam.org/growing_organic/3_advocacy_lobbying/eng_leaflet_PDF/Rural_Development.pdf`.

Kaganzi, E., Ferris, S., Barham, J., Abenakyo, A., Sanginga, P., Njuki, J., 2009. Sustaining linkages to high value markets through collective action in Uganda. Food Policy 34 (1), 23–30.

Kaplinsky, R., 2000. Globalisation and unequalisation: What can be learned from value chain analysis? Journal of Development Studies 37 (2), 117–146.

Kilian, B., Jones, C., Pratt, L., Villalobos, A., 2006. Is sustainable agriculture a viable strategy to improve farm income in Central America? A case study on coffee. Journal of Business Research 59 (3), 322–330.

Kramer, J. W., 2000. Co-operatives as Actors in a Market Economy. Berlin Cooperative Papers No. 50. Institut für Genossenschaftswesen an der Humboldt-Universität zu Berlin, Berlin.

Linton, A., 2008. A niche for sustainability? Fair labor and environmentally sound practices in the specialty coffee industry. Globalizations 5 (2), 231–245.

Magretta, J., 2002. Why business models matter. Harvard Business Review 80 (5), 86–92.

Markelova, H., Meinzen-Dick, R., Hellin, J., Dohrn, S., 2009. Collective action for smallholder market access. Food Policy 34 (1), 1–7.

Mutersbaugh, T., 2002. The number is the beast: A political economy of organic-coffee certification and producer unionism. Environment and Planning A 34, 1165–1184.

Neilson, J., 2008. Global private regulation and value-chain restructuring in Indonesian smallholder coffee systems. World Development 36 (9), 1607–1622.

Osterhaus, A. (Ed.), 2006. Business unusual. Successes and challenges of fair trade. FLO (Fairtrade Labeling Organizations International), IFAT (International Fair Trade Association), NEWS! (Network of European Worldshops) and EFTA (European Fair Trade Association), Newcastle-upon-Tyne.

Pelupessy, W., Díaz, R., 2008. Upgrading of lowland coffee in Central America. Agribusiness 24 (1), 119–140.

Philpott, S. M., Bichier, P., Rice, R., Greenberg, R., 2007. Field-testing ecological and economic benefits of coffee certification programs. Conservation Biology 21 (4), 975–985.

Pickard, D., 1970. Factors affecting success and failure in farmers co-operative organizations. Agricultural Economics 21 (1), 105–119.

Ponte, S., Ewert, J., 2009. Which way is "up" in upgrading? Trajectories of change in the value chain for South African wine. World Development 37 (10), 1637–1650.

Ponte, S., Gibbon, P., 2005. Quality standards, conventions and the governance of global value chains. Economy and Society 34, 1–31.

Rauch, P., 2007. Swot analyses and swot strategy formulation for forest owner cooperations in austria. European Journal of Forest Research 126 (3), 413–420.

Raynolds, L. T., 2000. Re-embedding global agriculture: The international organic and fair trade movements. Agriculture and Human Values 17 (3), 297–309.

Raynolds, L. T., Murray, D., Heller, A., 2007. Regulating sustainability in the coffee sector: A comparative analysis of third-party environmental and social certification initiatives. Agriculture and Human Values 24 (2), 147–163.

Raynolds, L. T., Murray, D., Taylor, P. L., 2004. Fair trade coffee: Building producer capacity via global networks. Journal of International Development 16 (8), 1109–1121.

Rigby, D., Cáceres, D., 2001. Organic farming and the sustainability of agricultural systems. Agricultural Systems 68 (1), 21–40.

Riisgaard, L., Bolwig, S., Ponte, S., Du Toit, A., Halberg, N., Matose, F., 2010. Integrating poverty and environmental concerns into value-chain analysis: A strategic framework and practical guide. Development Policy Review 28 (2), 195–216.

Shafer, S. M., Smith, H. J., Linder, J. C., 2005. The power of business models. Business Horizons 48 (3), 199–207.

Swinnen, J. F. M., Maertens, M., 2007. Globalization, privatization, and vertical coordination in food value chains in developing and transition countries. Agricultural Economics 37, 89–102.

Taylor, P. L., Murray, D. L., Raynolds, L. T., 2005. Keeping trade fair: Governance challenges in the fair trade coffee initiative. Sustainable Development 13 (3), 199–208.

Thorp, R., Stewart, F., Heyer, A., 2005. When and how far is group formation a route out of chronic poverty? World Development 33 (6), 907–920.

Valkila, J., 2009. Fair trade organic coffee production in Nicaragua – sustainable development or a poverty trap? Ecological Economics 68 (12), 3018–3025.

Willer, H., Yussefi, M. (Eds.), 2007. The world of organic agriculture. Statistics and emerging trends 2007, 9th Edition. International Federation of Organic Agriculture Movements (IFOAM) and Forschungsinstitut fuer biologischen Landbau (FiBL), Bonn.

Wills, C., 2006. Fair trade: What's it all about? In: Osterhaus, A. (Ed.), Business unusual. Successes and challenges of fair trade. FLO, IFAT, NEWS, and EFTA, Newcastle-upon-Tyne, pp. 7–27.

Wollni, M., Lee, D. R., Thies, J. E., 2010. Conservation agriculture, organic marketing, and collective action in the honduran hillsides. Agricultural Economics 41 (3-4), 373–384.

Wollni, M., Zeller, M., 2007. Do farmers benefit from participating in specialty markets and cooperatives? The case of coffee marketing in Costa Rica. Agricultural Economics 37 (2-3), 243–248.

Zerche, J., Schmale, I., Blome-Drees, J., 1998. Einführung in die Genossenschaftslehre: Genossenschaftstheorie und Genossenschaftsmanagement. Oldenbourg, München; Wien.

Chapter 7

The Added Value of Social and Environmental Standards for Smallholder Producers – A Case study of Nicaragua's Coffee Value Chains

TINA BEUCHELT, ANNA KIEMEN, THOMAS OBERTHUER AND MANFRED ZELLER

This chapter has been submitted in January 2011 as article for publication to the journal Agricultural and Human Values.

Abstract

Social and environmental certification schemes are production and trade standards which are increasingly promoted as an option for smallholder producers. As consumers pay higher prices for certified coffees, it is commonly assumed that, compared to conventional coffee, a higher share of the added value in consuming countries trickles down to the producers. Very few studies have addressed the difference between conventional and certified value chains, how power and information are distributed between different actors in the chain, and what amount of the added value created in consuming countries reaches the coffee producers. Based on the value chain concept, we analyze one conventional and two organic and Fairtrade certified value chain models with respect to their actors, governance structures, information flows, prices and producers' shares of final retail price. Quantitative and qualitative data were collected from 450 smallholder coffee producers and cooperative staff in Nicaragua, as well as from exporters, importers, and roasters. Results show that certified value chains tend to be more complex than conventional chains in both coffee producing and coffee consuming countries. The producers' share of final retail price in certified chains is substantially lower than in conventional chains (8-15% compared to 24-34%). Power is unequally distributed in all chains. Coffee prices, chain structure, and information flows differ between the cooperatives and their respective value chains. Using certification to add value does not benefit producers as much as expected due to information asymmetries, governance aspects, management skills, and complex chain structures causing high transaction costs.

Keywords:

Cooperative; governance; information asymmetry; price share; Organic-Fairtrade; value chain analysis

7.1. Introduction

Coffee sales are the main source of income for 25 million families on African, South Asian and Latin American hillsides; these farmers are mainly smallholders and the majority them are poor (Gresser and Tickel, 2002). As one solution to smallholders' poverty, social and environmental certification standards in coffee production are suggested. The oldest and best known social and environmental certification standards in coffee production are the fair trade and organic certifications (Raynolds, 2000). The fair trade movement was one of the earliest among the ethical and sustainable initiatives which trade directly with producers and aim at social justice and environmental well-being (Arnould et al., 2009; Ponte, 2002b). The ethical trade initiatives intend to counteract the dominant conventional commodity markets with their unfair price relations in production and trade (Raynolds, 2000).

While consumption of organic and fair trade certified products in industrialized countries first began in the 1970s, the products were only offered locally and represented a small market niche. The food safety crises and workers' rights scandals from the mid-1990s onwards brought environmental and ethical issues to the attention of a broad range of consumers (Codron et al., 2006). In the last decade, environmental and social attributes of food products and production processes have become increasingly important to consumers (Codron et al., 2006; Ponte and Gibbon, 2005; Raynolds, 2000; Rigby and Cáceres, 2001). Food products which are produced and processed according to environmental standards, like organic farming practices, have moved from niche market existence to mass markets in the past ten years. In part, this was due to the response of supermarket chains to the growing awareness and demands of consumers. With the introduction of fair trade labeled products in large supermarket chains and discounters, a similar movement from niche to mass markets is likely to occur for ethical and fair trade products (Codron et al., 2006). For example, certified coffee markets have experienced strong growth in Europe (Murray and Raynolds, 2006) and continue their development in the USA where growth rates of 56% for organic and 33% for Fairtrade[1] coffee are reported (Giovannucci and Villalobos, 2007). Fairtrade coffee consumption world wide is growing annually 20% (FLO, 2007).

Due to the increasing popularity of certified and gourmet coffees, many coffee producers and their organizations choose product differentiation to upgrade their coffee. Product differentiation enables farmers to gain a competitive advantage and obtain higher coffee prices. Investing in certification is one possible strategy for small-scale producers to

---

[1]The term 'fair trade' is written in lower case and two words where we refer to the movement of trading goods fairly. When the term 'Fairtrade' is capitalized we refer to one specific fair trade standard and label, the 'Fairtrade certification' of the *Fairtrade Labeling Organizations International* (FLO). Apart from the Fairtrade standard of FLO, there exist other fair trade standards developed by other certification agencies like the 'Fair for life' standard of the international *Institute for Marketecology* (IMO). Yet, during the time of research, other fair trade standards were not very common in smallholder coffee production.

upgrade and it enables smallholders to access niche markets based on environmental and social standards for production and processing (Pelupessy and Díaz, 2008; Ponte, 2002b; Riisgaard et al., 2010). In the coffee sector, production shortages and oversupplies are common leading to highly volatile coffee prices and frequent price crashes (Cashin et al., 2002). Because coffee is perennial and investment costs are high, short-term production adjustments in response to changes in market price are difficult; smallholders hope that participation in environmental and social certification schemes helps protect from the effects of market price fluctuations and adds more value to their coffee. Participation in certification schemes is supported by governments, NGOs, and donors as they see certification as a feasible way to increase income and buffer price volatility, thus reducing poverty among smallholders (Kilian et al., 2006; Neilson, 2008). Coffee certification is an important approach in Nicaragua where coffee is a major export product and source of income for smallholders. Like many smallholder coffee producers in Central America, Nicaragua's producers are mostly poor and were hit hard by the last coffee crisis from 1998/99 to 2002/03[2](Lewin et al., 2004).

Since the organic and Fairtrade certification standards have been explained in detail by several studies (Barrett et al., 2001; Browne et al., 2000; Mutersbaugh, 2002; Raynolds, 2000; Raynolds et al., 2007; Rigby and Cáceres, 2001; Tallontire, 2009), here they are just briefly described. The Fairtrade certification is an international private ethical standard[3] targeting social and trading aspects for producers in developing countries. It includes a guaranteed minimum price for coffee plus a premium for social investments, pre-financing of coffee on behalf of the buyers, and some basic environmental requirements. The Fairtrade certification aims for direct trade between farmers, cooperatives[4] and coffee importers/roasters (Raynolds, 2002), for increasing transparency in the coffee value chain and for enhancing benefits to democratically organized small producers (Wills, 2006). The main objective is for producers to sell directly to cooperatives, instead of to intermediaries who tend to offer lower prices (Wollni and Zeller, 2007). The cooperatives are then linked to wholesalers in coffee consuming countries through a single market intermediary.

Organic production standards are defined by both governments and civil society organizations. When organic products are imported by a country, as a minimum the national organic standards must be met. The organic production standards encompass sustainable resource management, natural inputs and maintenance of soil fertility, while prohibiting the use of synthetic agrochemicals (IFOAM, 2009a,b). As the Fairtrade standard holds

---

[2]During the coffee crisis, coffee prices fell below the production costs which led to the impoverishment of many smallholders worldwide (Lewin et al., 2004).

[3]The standards are set by the Fairtrade Labeling Organizations International (FLO). The control of the standard and the certification is issued by the FLO certification company FLO-Cert.

[4]In case cooperatives have no export license, exporters are permitted and can obtain a FLO certification.

many principles in common with organic certifiers, it is encouraging producers to convert to organic production (Renard, 2005).

Certified coffees are marketed at a higher retail price than conventional coffees. This price premium is generally assumed to compensate for higher production costs and/or lower yields inherent to organic farming and to give a 'fair' price covering production and living costs in the Fairtrade sector (Wills, 2006). The direct trading relationships in Fairtrade chains should lead to a higher share of the retail price being returned to the producer than in the conventional chain (Arnould et al., 2009). According to Slob (2006, p 134) the *"percentage of the final retail value that is retained by the producers is much higher in the Fair Trade system than in the mainstream market"*. Therefore, Fairtrade is a *"feasible alternative to the unfair distribution of value that characterizes today's mainstream coffee market"* (Slob, 2006, p 139). Organic movements also claim benefits to producers and rural development through the enhancement of economic, social and environmental sustainability (IFOAM, 2006). According to Fitter and Kaplinsky (2001), it is not necessarily given that the returns to differentiation are captured by producers even when they can meet the requirements of the differentiated markets. Regarding the claimed benefits of social and environmental certifications, research so far has focused on price effects for differentiated coffees at farm-gate or cooperative level. Results are not consistent, some studies show positive effects on prices (Arnould et al., 2009; Raynolds et al., 2004; Wollni and Zeller, 2007), others rather mixed or negligible effects (Kilian et al., 2006; Philpott et al., 2007).

Research on the distribution of value and price shares in certified agricultural value chains is scarce. There are only two studies which analyze the product upgrade to certified coffees regarding the share of the added value that reaches coffee producers (Daviron and Ponte, 2005; Mendoza and Bastiaensen, 2003). Mendoza and Bastiaensen (2003) found that, before the coffee crisis, conventional producers received a higher share of final retail price than Fairtrade certified producers (18% compared to 14% in 1996). This relation reversed during the coffee crisis when Arabica[5] coffee marketed via a Fairtrade chain for instant coffee in the UK captured a share of 11% of the final retail value compared to 7% in the conventional Nestlé chain (Mendoza and Bastiaensen, 2003). Daviron and Ponte (2005) also show that during the coffee crisis Ugandan conventional producers achieved shares of final retail price of 4% to 9% depending on quality and variety, organic producers less than 5% and Fairtrade certified producers 11.5% in a US value chain and 21% in an Italian value chain.

These two studies indicate that certified coffee producers do not necessarily receive much higher shares of final retail prices than conventional producers, but the Fairtrade certification seems to have a stabilizing function when conventional coffee prices are very

---

[5]Mainly two coffee species are commercially grown and traded globally, *Coffea Arabica* and *Coffea Canephora* (also known as Robusta). The coffee quality of *Coffea Arabica* beans is considered to be higher than of *Coffea Canephora* and thus, is traded at higher prices.

low, as occurred during the coffee crisis. Both studies focus on one certification standard at the time and so far, double certifications have not been analyzed. Although it is unknown whether the double certification contributes to higher prices and price shares for smallholders than only one certification, the double certification Organic-Fairtrade is increasingly demanded by coffee buyers and consumers, burdening producers with higher certification and investments costs.

The question is not only what share of the final retail price the coffee producers receive, but also how power is distributed and what information is communicated in the value chain. Both are determining factors for the successful integration of smallholder producers into the value chain (Gereffi et al., 2005; Talbot, 2002). Power and information issues have been overlooked in research on certified coffee value chains (Tallontire, 2009).

At present, only weak empirical evidence is available to international donors and governments as the basis for their support to coffee certification schemes as tools which link farmers to high-value markets, increase producers' share of final retail price, and thus contribute to poverty reduction. The analysis of certified (coffee) value chains as compared to conventional value chains has hardly been addressed by research.

However, to guide policy-makers in decisions that foster better market integration of, and increase economic benefits to, smallholder producers, more information on the added value of organic and fair trade certifications as well as changes in power and information relations between buyers and sellers are of great importance.

This paper contributes to filling the knowledge gap with an analysis of Fairtrade and Organic-Fairtrade certified value chains compared to a typical conventional coffee value chain. Two research questions are explored: first, what are the structures of the Fairtrade and Organic-Fairtrade certified chains compared to a conventional coffee value chain in terms of actors, governance, and information flows? And second, what is the effect of adding value to coffee through Fairtrade and Organic-Fairtrade certification on farm-gate prices and the producers' share of the final retail price?

The article is structured in six sections. The next section provides a brief overview of governance and upgrading in coffee value chains. It is followed by a description of the research methodology. In the fourth section, three different value chain models of conventional, Fairtrade and Organic-Fairtrade certified cooperatives are described and compared regarding their actors, governance, and information flows. The fifth section analyzes the value added by upgrading through certification at the producers' level in terms of prices and price shares. The last section contains the conclusions.

7.2. Upgrading and governance in coffee value chains

A value chain is comprised of all necessary activities to create and move a good or service along the different steps of design, production, marketing, consumption, and disposal after use (Kaplinsky, 2000). It links the different actors in the chain which are characterized by buyer and supplier relationships (Bolwig et al., 2010). The coffee value chain is split in two: coffee producing countries, mostly developing countries with millions of smallholders involved, and consuming countries, mainly industrialized countries with comparatively rich consumers (Ponte, 2002a). In coffee producing countries, the producers' value captured for conventional coffee production and primary processing over the past 30 years has varied between 10% and 20% of the final retail value with a peak in 1975/76 when producers received a 30% share (Daviron and Ponte, 2005; Fitter and Kaplinsky, 2001; Talbot, 1997). The share declined in the 1990s to less than 14% with even lower shares during the coffee crisis at the turn of the millennium (Daviron and Ponte, 2005). In coffee consuming countries import and trade activity adds up to 40% of the final retail price; roasting adds a further 20%. In the 1970s and 1980s, the total value captured in consuming countries was between 40% and 50%; by the end of the eighties, this share had risen to over 70% (Daviron and Ponte, 2005; Talbot, 1997).

Kaplinsky (2000) notes that, in general, the gains in the value chains are increasingly found in areas outside production. While there are many contributing factors which lead to a decline in producers' share of total income in developing countries, one important issue is low entry barriers for the production of commodity products which leads to high competition and reduces production profits (Kaplinsky, 2000). Large benefits due to differential productivity of factors and barriers to entry are increasingly derived by design and branding activities in the value chains (Kaplinsky, 2000). In the coffee value chain, design and branding usually takes place in the coffee consuming countries, thus most value adding is realized in industrialized countries (Daviron and Ponte, 2005).

Differentiation through upgrading can be understood as the move towards more (economically) rewarding functional positions in the value chain, the production of goods which provide better returns and have more added value, and/or rewards through reducing risks (Ponte and Ewert, 2009; Riisgaard et al., 2010). Product upgrading is the change in production to increase the unit value of goods by, for example, complying with buyer requirements for quality, environmental and social standards (like the organic or Fairtrade certifications) (Riisgaard et al., 2010).

The governance structures in value chains can support but also constrain upgrading activities. For producers in developing countries, the type of governance determines their market access, their function in the chain and upgrading possibilities, the availability of information and technology, attainment of knowledge, and their possible gains. Governance in global value chains can be understood as the organization of activities within the chain; who does what in a value chain is determined by the power relationships between actors. Some actors have the power to define rules and standards, to include or exclude chain participants, to decide upon the value-adding activities of each participant, and to monitor the performance of other actors (Fitter and Kaplinsky, 2001; Ponte and Gibbon, 2005). Governance is the process of organizing activities to achieve a functional division of labour along the value chain which leads to specific resource allocations and distributions of gains

(Ponte and Gibbon, 2005). In the commodity coffee chain, governance aspects are not as important as in gourmet and certified coffee chains (Fitter and Kaplinsky, 2001), but even the commodity coffee value chain is determined by increasing buyer power (Ponte, 2002a). Therefore, coffee value chains are typically considered to be buyer-driven chains (*ibid.*), led by retailers, branded marketers, and industrial processors like roasters (Ponte and Gibbon, 2005). Especially the roasters maintain a dominant position in the commodity chain as they effectively manage information asymmetry on quality (*ibid.*).

7.3. Methodology

The field research was conducted in 2007 and 2008 in three important coffee-producing provinces of northern and central Nicaragua: Matagalpa, Madriz, and Nueva Segovia. One conventional and two Fairtrade and Organic-Fairtrade certified coffee cooperatives with their respective value chains were purposely selected from cooperatives participating in a larger overarching research project. Selection criteria for the cooperatives were certification type and the willingness of cooperatives as well as their coffee buyers to share sensitive business information. Coffee producers were associated with first-level cooperatives. In the case of the certified value chains, the first-level cooperatives formed umbrella organizations called second-level cooperatives, which usually assume marketing and extension responsibilities. In 2008, semi-structured interviews were conducted with 34 smallholder coffee producers in 22 communities, seven presidents of first-level cooperatives, five second-level cooperative staff members, one coffee exporter, a German importer, a German direct importing roaster and one US roaster. Additionally, three focus group workshops for each chain model were conducted with a total of 31 participants.

As cooperative selection and complementing quantitative data is based on the overarching research project, its survey method is now briefly described. Six cooperatives were randomly selected based on a list containing all existing smallholder cooperatives in the region. Depending on the cooperative, either a simple random sampling method or a two-stage random sampling method was applied to draw the cooperative members for the household survey. In total, 327 cooperative members were surveyed with a structured questionnaire. The quantitative data was complemented by qualitative data from 48 key-person interviews with cooperative staff, exporters, roasters, and researchers, as well as 33 semi-structured producer interviews and 21 focus group discussions.

We distinguish three value chain models in which the selected cooperatives operate. The value chain model of the conventional cooperative is called Model C. The value chain models of the cooperatives are called Model A, for the Organic-Fairtrade certified cooperative, and Model B for the conventional Fairtrade and Organic-Fairtrade certified cooperative.

7.4. Comparison of the conventional and certified coffee value chain models

This section first describes the three value chain models. Then, the differences in the value chain models are assessed with respect to actors, governance, and information flows.

7.4.1. The conventional coffee value chain – Model C

The conventional cooperative is a multifunctional cooperative with around 300 members, of which the majority are coffee producers. The cooperative does not buy coffee directly from its members and is not involved in coffee storage or processing. It has a strategic alliance with an exporter to market its members' coffee, based on a long-term, trustful relationship. The exporter is a subsidiary of an international trading company and one of the two major exporting companies in Nicaragua. The exporter/trader provides credit and pre-financing of the harvest to the cooperative for a contracted coffee quantity to be delivered by the cooperative's members (Figure 7.1). Based on this credit, the coop-

Figure 7.1.: Simplified structure of the conventional value chain – Model C

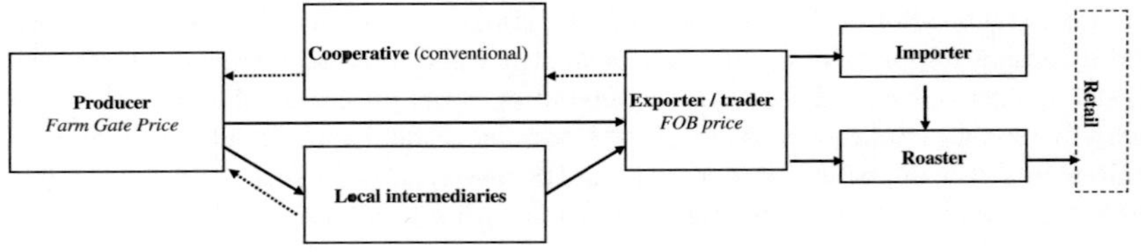

Note: The value chain beyond the exporter is based on information provided by the exporter and could not be further verified. Yet, it is like the typical chain structure in the conventional coffee chain (Ponte, 2002a).
Source: Own illustration.

erative can provide credit and pre-financing of the harvest to its members. At harvest time, producers deliver the coffee directly to the exporter but receive payment from their cooperative. Through decoupling coffee delivery from payment, the cooperative maintains the guarantee that their credits are paid back. Producers are paid a local price, which is influenced by the coffee price at the New York stock exchange (now the Inter-continentalExchange) and by the average regional coffee quality. The exporter blends the received coffee with other coffees according to the buyer's requirements and sells in bulk. Therefore, no direct links to chain actors in importing countries could be identified, but the exporter stated that, in general, the buyers are large US and European importing and roasting companies. The value chain via the exporter is short as there are few intermediaries and traders involved, in both the coffee producing and consuming countries. Farmers also sell around 43% of their coffee to different local intermediaries who sell on to one of the two major exporters in Nicaragua. The coffee sales outside the cooperative are partly due to paying back informal credits and partly due to temporarily higher prices offered by the competing major exporter.

7.4.2. Certified value chain – Model A

The certified value chain Model A is based on a second-level cooperative composed of six first-level cooperatives with a total of around 400 members. The second-level cooperative administers the Fairtrade and organic certification. Nearly all producers have certified organic coffee, only some are in the transition period of becoming organic certified. The producers are members of the first-level cooperatives, but receive support like extension or credits directly from the second-level cooperative. The first-level cooperatives' main responsibility is representing the producers' interests regarding the use of the Fairtrade social premium for social investments. The second-level cooperative owns a dry processing mill with a cupping laboratory for quality check. It processes the members' coffee and evaluates samples from each member to identify individual quality (physical and sensory). It has an export license and actively exports.

At harvest time, producers can choose to either receive an immediate payment at spot market price upon coffee delivery to the second-level cooperative or wait for the final payment in April or May. The final payment is done a couple of weeks after the second-level cooperative has finished coffee sales and accounting. In general, farmers opt for the later payment with higher prices. Yet, to either have immediate cash or pay back their informal credits farmers sell around 28% of their harvest not to the second-level cooperative but to other intermediaries (Figure 7.2). The second-level cooperative sells only the coffee in transition (around 3%) to an exporter at a price equivalent to the Fairtrade minimum price or higher. All Organic-Fairtrade coffee is sold to long-term partners in Germany and the USA. In the US-chain, the specialty roasting company contracts with the cooperative but relies on the logistical support from an importer. In Germany, one company imports and labels the coffee while subcontracting for the roasting process; the other imports and sells the coffee to different non-profit enterprises or organizations which roast and label it under their own brands. The American roasting company paid an extra premium of around 0.10US\$/lb for high quality[6] in addition to the Fairtrade minimum price and the organic and social premiums. The additional premium for high quality is equally distributed to all producers and not according to the individual producers' quality results. Due to a different management and trading approach, the German importers and roasters do not pay extra premiums for high quality (see subsection 7.4.4).

Figure 7.2.: Simplified structure of the certified value chain – Model A

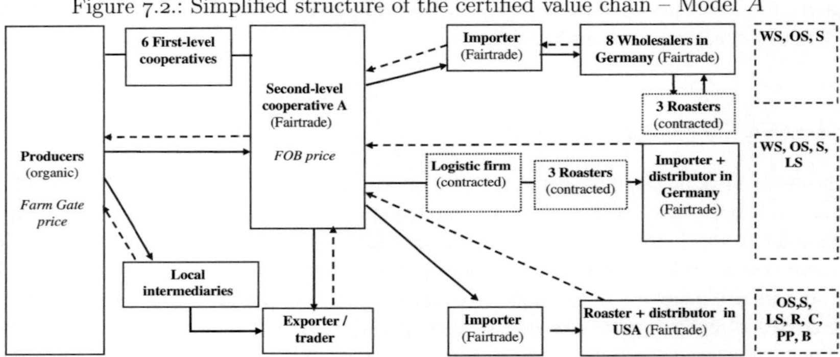

Note: WS One-world Shops, OS Organic Stores, S Supermarkets, LS Large Scale Customer, R Restaurants, C Convenience stores, PP Private and Public Institutions (such as Universities and Offices), B Bakery and Coffee Shops.
Source: Own illustration.

7.4.3. Certified value chain – Model B

The certified chain Model B is also a Fairtrade certified second-level cooperative whose members are eleven first-level cooperatives and two cooperative unions with twelve fur-

---

[6]In this case, high quality is determined by the roaster based on the cupping result which follows the grading scheme for coffee from the Specialty Coffee Association of America. As the cooperative owns a cupping laboratory, they apply the same grading scheme as the roaster does. Thus, they know the quality of their coffee which enhances their bargaining power on prices.

ther affiliated first-level cooperatives (Figure 7.3). Altogether, more than 2,000 coffee producers are united in the second-level cooperative. In contrast to the certified Model *A*, some of the first-level cooperatives and cooperative unions support farmers directly with extension, pre-financing and coffee procurement. The second-level cooperative owns a dry mill and an export license, and is processing and cupping the coffee.

Figure 7.3.: Simplified structure of the Fairtrade certified coffee value chain – Model *B*

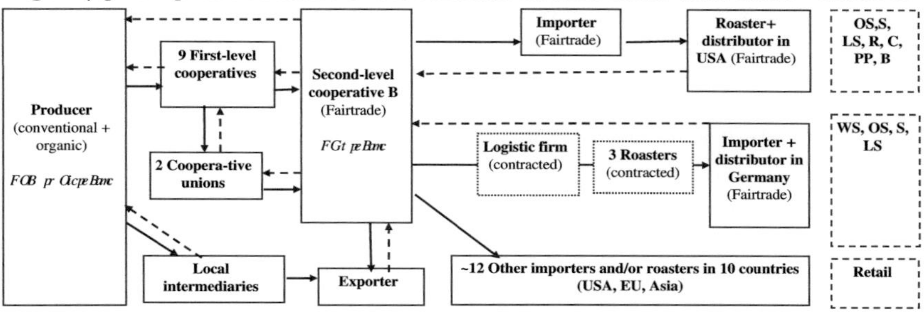

→ Coffee delivery (physical product flow)

⋯▶ Credit provided to chain actors (pre-financing of the contracted coffee quantity)

Note: WS One-world Shops, OS Organic Stores, S Supermarkets, LS Large Scale Customer, R Restaurants, C Convenience stores, PP Private and Public Institutions (such as Universities and Offices), B Bakery and Coffee Shops.
Source: Own illustration.

In this chain model, 40% of the coffee quantity is produced according to organic standards, less than 5% according to the standard of UTZ certified[7], and the remaining coffee is produced conventionally. Farmers also market their coffee outside their cooperative, exact figures are not available but there is an estimated amount of at least 25% off-sales. Despite the Fairtrade certification for all coffee producers, the second-level cooperative sells only 41% of the conventional coffee as certified Fairtrade coffee. Of the organic coffee, 70% is sold as Organic-Fairtrade. Although the demand by buyers was higher, the other 30% did not fulfill the quality standards for that marketing channel.

---

[7]According to the homepage of 'UTZ certified', the term 'utz' means good in a Mayan language (`http://www.utzcertified.org`, accessed 23 June 2010). It is a certification standard including a set of social and environmental criteria for coffee production and traceability regulations.

Upon coffee delivery, producers are paid local spot market prices at the daily rate. In the case of organic producers, the second-level cooperative supposedly adds a premium for organic coffee. Yet, in the last two years, the organic producers stated that they have not been paid higher prices than the local spot market price for conventional coffee. The Fairtrade social premium is not paid to the farmers but used by the cooperatives to invest in social and infrastructural projects, for example scholarships. The second-level cooperative sells at world market coffee prices and, depending on the chain they are selling to, receive organic or Fairtrade premiums. Coffee is sold to a dozen buyers ranging from a conventional exporter to alternative trade organizations, specialized intermediate traders and specialty roasters. Between most, long-term relationships exist. We interviewed only selected buyers choosing the same buyers as in chain Model A, so a comparison of the two chains is possible. Both buyers together make up 12% of the cooperative's traded coffee volume. The interviews and secondary data research indicated that the value chains of the other buyers look similar to the model presented here.

7.4.4. Differences in the three value chain models with respect to actors, governance and information flows

This subsection analyzes the differences and similarities in the three value chain models with respect to actors, governance, and information flows. They are summarized in Table 7.1. In the conventional chain Model C, the number of actors beyond the farm level is limited; there are the cooperative as credit and service provider, the exporter/trader, importers and roasters. It is a short and relatively direct value chain although the coffee cannot be traced to its final destination. The set up in the two certified chain models is more complex. The certified Model A has a similar number of producers as the conventional model and manages the coffee export itself. Compared to the other two chains, the certified Model B has not only more producers but also more actors in the first part of the value chain due to the various cooperative associations. Being members in first and second-level cooperatives and in cooperative unions at the same time reduces the identification of farmers with the second-level cooperative; members indicated that they do not feel like owners of the second-level cooperative or its possessions, like the dry mill. It also raises transaction costs for farmers and the various cooperative types because of the greater need for communication and coordination. The second-level cooperative in Model B has more buyers than there are in Model A. The cooperative must adjust to the diversity of the buyers which aim to please different market niches. There are also slight differences in the German and US roasting and marketing channels in the certified chains. The buyers in Germany are rather small and specialized. They subcontract certain services like logistics and roasting. In our case study, the US roaster is a very large company which also subcontracts import logistics. Yet in the certified US value chains, many small and specialized actors are present just as they are in Germany. In general, a shortening

Table 7.1.: Summary of the value chain models regarding actors, governance and information flows

	Conventional Model C	**Certified Model** A	**Certified Model** B
Actors			
In producing country	Limited number	Limited number	Few to several
In consuming country	Limited number	Few to many actors, often of small sized	Few to many actors, often of small sized
Governance			
Producers	No bargaining power on prices	No bargaining power on prices	No bargaining power on prices
Cooperatives (Coop.)	Currently no bargaining power on prices, but possible when differentiating; Long-term relationship with buyer (exporter)	Limited bargaining power; must adjust to certification demands of buyer; Long-term relationship with buyers but only yearly contracts	Limited bargaining power; must adjust to certification demands of buyer; Long-term relationship with buyers but only yearly contracts
Information flows			
Producers-Coop.	Medium information asymmetries	Low information asymmetries	Strong information asymmetries
Coop.-Buyers	Strong information asymmetries	Low to medium information asymmetries	Low to medium information asymmetries

of the certified value chains compared to the conventional chain is not observed in either the producing or the consuming countries. An important difference between our results and those of other studies is that we compared the certified chain to a conventional chain model where farmers are also cooperatively organized – instead of comparing it to unorganized farmers. As the conventional cooperative also offers marketing services via the exporter/trader, the main benefit in reducing the number of intermediaries in producing countries is not the certification but the organization of producers.

Governance in the coffee value chain refers to the power relationships between the buyers and the sellers. In the conventional chain, the cooperative's manager stated that she has no bargaining power on prices as they are set by the exporter and she depends on him due to the provided credits. In the future, the cooperative's manager sees a potential for bargaining as she gains knowledge about management and coffee marketing through voluntary higher education and plans to introduce a coffee differentiation strategy to the cooperative. The exporter has expressed the possibility of negotiating prices and supports the cooperative in its pursuit of coffee differentiation. The producers' position in price negotiation is not better in certified chains than in the conventional chain because the amount of premiums is determined by the buyers and the cooperatives. Cooperatives in the certified chains participate to a certain degree in the price negotiation process as roasters and importers demand a higher quality than the conventional market. Despite this, the cooperatives are dependent on buyers for pre-harvest financing due to the limited market possibilities for certified coffees. In order to keep the buyers, cooperatives must follow the buyer's specific certification requirements which can include the change from one organic standard to another, or switching certification agencies. Cooperative staff also indicated that quality is not always paid for. This can be due to the small size of some buyers and/or the complex chains of certified coffee with many actors in consuming countries which raises transaction costs; power and information asymmetries can also be assumed. Although long-term relationships between buyers and sellers exist in all coffee value chains, the contracts themselves are short-term and renegotiated annually.

In terms of information flows, producers in the conventional chain are in contact with cooperative staff through the extensionists, producers' visits to the cooperative office for credit facilitation, and the annual assembly. The conventional cooperative manager frequently communicates with the exporter about production and credits. Almost no information regarding prices, product attributes, and destinations for conventional coffee is released to the cooperative's manager or producers. In the certified chains, producers receive information about coffee, quality issues, and pricing at the annual general assembly, in monthly group meetings, and during monthly on-farm visits. Additionally, monthly meetings are organized by the leaders of first-level cooperatives to facilitate dialogue with the second-level cooperative.

Conventional producers know hardly anything about certification standards. Knowledge about organic production standards is usually communicated by cooperative staff in chains A and B. The system of Fairtrade remains unclear to nearly all producers and even to leaders of the first-level cooperatives in both certified chain Models A and B. The low knowledge levels regarding Fairtrade can be explained by the complexity of coffee marketing and the Fairtrade pricing and premium system. Most producers never finished primary school so their understanding of such information is at times limited. Furthermore, in Model B the cooperatives and their staff do not always have the time and/or interest to convey all the necessary information to farmers, creating more pronounced power asymmetries. Between the two certified chains, the quality and frequency of information flows varies considerably. In Model A, a high trust relationship between the individual producers and cooperative staff exists. In Model B, producers and presidents of first-level cooperatives stated that they are not satisfied with the relationship and do not trust their second-level cooperative, mainly because of the lack of transparency in the management of processing, payment, and relevant marketing information. The good information flows and relationships in Model A are certainly supported by the small number of cooperative members while the large membership of the second-level cooperative in Model B increases communication logistics and costs which hinder information flows. Additionally, personal commitment seems important. The manager of the second-level cooperative in Model A considers it as essential to have direct contact with the producers and to maintain personal relationships with members. In Model B, only the extension workers and the staff from the first-level cooperative, respectively the cooperative union, have direct contact with the producers. Communication is not good between the different cooperative levels and farmers and the roles of each organizational level are not clarified. There is a considerable alienation and loss of identification between the second-level cooperative and its coffee producing members; members are dissatisfied with the provided services. The information flows between the cooperatives and the buyers in the certified chain vary among the buyers; some buyers communicate little while others are very open and inform the cooperatives well. Yet, even the Fairtrade buyers do not share sensitive business information like profit margins.

7.5. The added value of certification and producers' shares of retail prices

One of the main incentives for producers and cooperatives to participate in certified markets is the assumed higher coffee price compared to conventional markets. When conventional coffee prices are good, as they were for the production years 2006/07 and 2007/08, producers do not necessarily receive higher prices for being members of a Fairtrade certified cooperative (Table 7.2). In 2007/08, the conventional producers in the Fairtrade Model B received a lower price than their non-certified colleagues in Model C. Double certification, Organic-Fairtrade, is also no guarantee for better prices. According to Fairtrade standards for the production year 2007/08, an organic premium of 0.15US\$/lb coffee must be paid in addition to the market price. The organic farm-gate prices for Model A are 0.14US\$/lb higher compared to Model C. The organic farm-gate prices in Model B are also 0.14US\$/lb higher than the conventional price offered within Model B, but there is only a 0.01US\$/lb difference compared to Model C. In Model A, the Organic-Fairtrade certified producers gain 0.23US\$/lb, or 21%, more than their colleagues in Model B. The cooperative of Model B stated that it was not able to distribute a premium due to missing gains from export. It also has low market shares for its conventional Fairtrade coffee. Reasons for both are likely a mix of lacking access to certified markets, low management skills, and low quality of coffee. A lower coffee quality alone is not a sufficient reason as the second-level cooperative also sells to the US specialty roaster. Large overhead and transaction costs in processing and administration of the second-level cooperative in Model B further contribute to lower prices.

Coffee farmers in the conventional chain receive a much higher share of the final retail price in the USA and Germany than farmers in the Fairtrade and Organic-Fairtrade chains. In 2006/07, conventional producers' shares of final retail price varied between 18% in the

Table 7.2.: Coffee prices and producers' share of retail price in the three coffee value chain models, coffee harvest 2007/08

	Farm-gate[a]	FOB[b]	Retail[c]	Producers' retail share
	(US\$/lb)	(US\$/lb)	(US\$/lb)	(%)
Conventional chain – Model C				
Farmer - Cooperative/exporter - Retail Germany	1.07	1.31	4.31	24.83
Farmer - Cooperative/exporter - Retail USA	1.07	1.31	3.12	34.30
Certified chain – Model A				
Farmer - 2nd-level Cooperative - Importer/Distributor - Retail Germany	1.31	1.78	10.52	12.45
Farmer - 2nd-level Cooperative - Importer - Distributor - Retail Germany	1.31	1.78	8.47	15.47
Farmer - 2nd-level Cooperative - Importer - Roaster in USA	1.31	1.87	11.41[d]	11.48
Certified chain – Model B				
Farmer - Conventional 1st-level Cooperative - 2nd-level Cooperative - Importer - Roaster in USA	0.94	1.58	11.41[d]	8.24
Farmer - Organic 1st-level Cooperative - 2nd-level Cooperative - Importer - Roaster in USA	1.08	1.84	11.41[d]	9.47
Farmer - Organic 1st-level Cooperative - 2nd-level Cooperative - Importer/Distributor I - Retail Germany	1.08	1.84	10.52	10.27

Note: [a] The farm-gate price is lower than the 'Free on Board (FOB)' price the cooperative receives as costs for processing, export, and administration are subtracted. The cooperative's management, the product's quality and buyers demand further influence how much and at what price coffee can be sold as certified coffee. The producers obtain a mixed price averaged across all market channels of the cooperative.
[b] Converted to green exportable coffee.
[c] Green coffee equivalent price. Conversion factor for roasted/green coffee=0.84. VAT included. Retail price in Model C is based on secondary data, not on own data, and was adjusted for the USA with the consumer price index for roasted coffee according to the Bureau of Labor Statistics for 2008, 2010, http://www.bls.gov/cpi/cpid08av.pdf (accessed 16 April 2010).
[d] The roaster did not distinguish in retail price between conventional and Organic-Fairtrade coffee.
Source: Own data (2008) and ICO (2009).

German and 24% in the US retail market. The shares of conventional producers for the year 2007/08 are 24% and 34%, respectively. These shares are higher than the 10% to 20% literature usually indicates, but still within the range of past peaks (Daviron and Ponte, 2005; Fitter and Kaplinsky, 2001). In the Fairtrade chains, conventional producers had a share of the final retail price of 8% and organic producers of 9% to 15% in 2007/08. These shares are similar to those obtained before and during the coffee crisis. The producers in Model A achieved higher proportions of retail price than those in Model B, although the difference is not as large as the variation in the Organic-Fairtrade farm-gate prices suggests. The price shares indicate that, as opposed to the conventional producers, certified producers have not been able to capture much higher shares of the retail price since the coffee crises.

Several aspects explain the differences between the conventional and the certified farm-gate prices and producers' shares of final retail price. First, the certified cooperatives' management pointed out that increasing world market prices lead to lower price premiums for organic and Fairtrade coffees in comparison to conventional coffees. Therefore, the producers' share of retail price is influenced by the conventional coffee prices. This is supported by a study of Bettendorf and Verboven (2000) showing that higher green bean coffee prices do not translate to equally higher retail coffee prices in the Netherlands because raw material costs are only a small part of the whole processing and marketing costs.

Second, cooperatives are not always able to market all their coffee as certified coffee (like in certified Model B) and thus, the farm-gate coffee price is based on a mixed calculation of the different prices obtained from each marketing channel. The varying market shares are due to the increased international competition, the relatively low demand for certified coffees in relation to world coffee demand, quality issues, and marketing skills of the cooperative staff.

Third, the certified cooperatives who export must process the coffee into an export-ready green bean by drying, sorting, husking, packing, and organizing administration, transport, and export. The interviews showed that the cooperatives' costs for these activities are around 0.10US\$/lb coffee higher than those of the exporter. The exporter probably benefits from economies of scale leading to lower processing and marketing costs. This could also explain why conventional prices in Model C were higher than those in Model B. While Fairtrade advocators assume that increased control of processing and export activities in the value chain leads to higher income for producers and their organizations (Slob, 2006), our data indicate that it is not necessarily true. In fact, it would be cost-saving to pay an exporter for these services, but this decision needs to be weighed against the expense of increasing dependency.

Fourth, the distribution channels in the conventional market are dominated by a few, very large importing and roasting companies (Pelupessy and Díaz, 2008). They possess economies of scale and thereby reduce transaction costs. This is more difficult in the alternative trading and retail models of smaller certified buyers and roasters. Another explanation was added by the executive manager of the non-profit Fairtrade importer/distributor in Germany. He mentioned that they deliberately do not pay a higher price for quality than the Fairtrade standard demands. The extra margin from sales of the higher quality coffee is used to cover losses caused by new cooperatives which have no trade experience in international coffee markets and raise transaction costs for the enterprise during their learning process. According to the executive manager, the enterprise aims to enable inexperienced cooperatives to participate in the certified coffee market and thus to contribute to local development. The objective is fulfilled when smoothly run cooperatives start selling to other buyers. This occurred in Model A where the German enterprise was the cooperative's first buyer. With increasing experience, the cooperative has switched to the US roaster which has higher professional and quality demands but also offers better prices.

7.6. Conclusions

Three value chain models of conventional coffee producers and producers who upgraded through social and environmental coffee certification were analyzed with respect to their actors, governance, information flows, prices, and achieved price shares. Certified value chains tend to have more and smaller-sized actors, especially in consuming countries, compared to the conventional chain. Farmers are found to have no bargaining power over prices irrespective of the value chain, while certified cooperatives have limited bargaining power towards their buyers compared to the cooperative conventional chain. Power is unequally distributed between buyers and sellers of coffee in all chains. The quantity and quality of information flows depends on the cooperative and value chain model. Information asymmetries are fewer in certified chains, yet also depending on the cooperative. Organic-Fairtrade coffee prices can be higher than conventional prices, but the price difference and market shares of certified coffee vary according to the cooperative. The smaller second-level cooperative with fewer intermediating cooperative types of Model A showed better results in both aspects.

The producers' share of the final retail price is substantially lower in the Fairtrade and Organic-Fairtrade certified chains than in the conventional chain. This result is contrary to the claims of Fairtrade advocacy. One explanation is that the producers' share of retail price is influenced by the conventional coffee prices, when they are good producers in certified chains capture lower shares of the final retail price than when conventional coffee prices are bad. The Fairtrade certification does not necessarily change the unfair distribution of value that characterizes the mainstream coffee market.

The value added through certification can easily evaporate at the cost of the primary producer when certified value chains are characterized by more and smaller-sized chain actors in consuming countries, information and power asymmetries as well as governance problems in intermediating cooperatives exist. In the case of Nicaragua's central and northern coffee producers, product upgrading through participation in certification schemes is not as beneficial regarding prices and value distribution as assumed by governments, NGOs, and international donors.

We conclude that the organization of producers as such, but also their organizational structure, capacity, and efficiency of the organization is important for successful coffee marketing. The conventional value chain should be reconsidered by researchers, policy makers, and NGOs as a system which – with the right chain structure – could also benefit producers due to higher efficiencies and lower transaction costs. Further research on the value chain could identify the determinants for successful market and value chain integration of producer organizations, and investigate where resources could be pooled to reduce transaction costs and raise efficiencies, not only in producing but also in coffee consuming countries.

In order to better link farmers to (high-value) markets and to increase their income, it is recommended to focus more on the structure and functioning of producer organizations and their respective value chains, analyzing the actual bottlenecks in marketing. In the coffee producing countries, it is recommended to improve the organizational and marketing skills of cooperative staff, enabling them to exercise more bargaining power in the long-term. It is important to increase the efficiencies of cooperatives; this could also include the downgrading of some functions taken in the value chain like the processing and exporting of coffee. Possibilities to increase cooperation with exporters should be further explored as they can play a crucial role through their expertise, financial resources, and economies of scale. In order to increase the producers' price and improve the distribution of value shares, additional attention should be given to enhancing trade, processing, and marketing efficiencies in certified value chains in consuming countries. This could be a reorganization of the alternative trade sector with its many small profit or non-profit enterprises and organizations processing and marketing fair trade coffee. These actors could consolidate to exert economies of scale and thus reduce the transaction costs occurring in consuming countries. While this is certainly a new way of thinking in the alternative trade sector, streamlining could effectively contribute to its mandate of improving the marketing and trading conditions of poor smallholders, and thus fostering the socio-economic development of disadvantaged producers.

Acknowledgements

It is a pleasure to thank Paul-Theodor Schütz for his valuable comments. The funding for field research by the *Beratungsgruppe für entwicklungsangewandte Agarforschung* (BEAF) of the *Deutsche Gesellschaft für Technische Zusammenarbeit* (GTZ) and the logistic support by the *International Center for Tropical Agriculture* (CIAT) in Colombia and Nicaragua, especially by Axel Schmidt, is gratefully acknowledged.

7.7. Bibliography

Arnould, E. J., Plastina, A., Ball, D., 2009. Does fair trade deliver on its core value proposition? Effects on income, educational attainment, and health in three countries. Journal of Public Policy & Marketing 28 (2), 186–201.

Barrett, H. R., Browne, A. W., Harris, P. J. C., Cadoret, K., 2001. Smallholder farmers and organic certification: Accessing the EU market from the developing world. Biological Agriculture and Horticulture 19 (2), 183–199.

Bettendorf, L., Verboven, F., 2000. Incomplete transmission of coffee bean prices: Evidence from The Netherlands. European Review of Agricultural Economics 27 (1), 1–16.

Bolwig, S., Ponte, S., Du Toit, A., Riisgaard, L., Halberg, N., 2010. Integrating poverty and environmental concerns into value-chain analysis: A conceptual framework. Development Policy Review 28 (2), 173–194.

Browne, A. W., Harris, P. J. C., Hofny-Collins, A. H., Pasiecznik, N., Wallace, R. R., 2000. Organic production and ethical trade: Definition, practice and links. Food Policy 25 (1), 69–89.

Cashin, P., McDermott, C. J., Scott, A., 2002. Booms and slumps in world commodity prices. Journal of Development Economics 69 (1), 277–296.

Codron, J.-M., Siriex, L., Reardon, T., 2006. Social and environmental attributes of food products in an emerging mass market: Challenges of signaling and consumer perception, with European illustrations. Agriculture and Human Values 23 (3), 283–297.

Daviron, B., Ponte, S., 2005. The coffee paradox. Global markets, commodity trade and the elusive promise of development. Zed Books, London and New York.

Fitter, R., Kaplinsky, R., 2001. Who gains from product rents as the coffee market becomes more differentiated? A value-chain analysis. IDS Bulletin 32 (3), 69–82.

FLO, 2007. Annual report 2007. An inspiration for change. Fair Trade Labelling Organizations International (FLO), Bonn, accessed 15 December 2010.
URL `http://www.fairtrade.at/pics/infomaterial/FLO_AR2007.pdf`.

Gereffi, G., Humphrey, J., Sturgeon, T., 2005. The governance of global value chains. Review of International Political Economy 12 (1), 78–104.

Giovannucci, D., Villalobos, A., 2007. The state of organic coffee: 2007 US update. Sustainable Markets Intelligence Center (CIMS), San Jose, accessed 15 December 2010.
URL `http://orgprints.org/12856/1/State_of_Organic_Coffee_2007_US_update.pdf`

Gresser, C., Tickel, S., 2002. Mugged: Poverty in your cup. Oxfam International; Make Trade Fair, Washington D.C., accessed 21 November 2006.
URL `http://www.oxfamamerica.org/newsandpublications/publications/research_reports/mugged`.

IFOAM, 2006. Organic agriculture and rural development. International Federation of Organic Agriculture Movements (IFOAM), Bonn, accessed 15 December 2010.
URL `http://www.ifoam.org/growing_organic/3_advocacy_lobbying/eng_leaflet_PDF/Rural_Development.pdf`.

IFOAM, 2009a. Definition of organic agriculture. International Federation of Organic Agriculture Movements (IFOAM), Bonn, accessed 29 March 2010.
URL `http://www.ifoam.org/growing_organic/definitions/doa/index.html`.

IFOAM, 2009b. The principles of organic agriculture. International Federation of Organic Agriculture Movements (IFOAM), Bonn, accessed 15 December 2010.
URL `http://www.ifoam.org/about_ifoam/principles/index.html`.

Kaplinsky, R., 2000. Globalisation and unequalisation: What can be learned from value chain analysis? Journal of Development Studies 37 (2), 117–146.

Kilian, B., Jones, C., Pratt, L., Villalobos, A., 2006. Is sustainable agriculture a viable strategy to improve farm income in Central America? A case study on coffee. Journal of Business Research 59 (3), 322–330.

Lewin, B., Giovannucci, D., Varangis, P., 2004. Coffee markets: New paradigms in global supply and demand. Agriculture and Rural Development Discussion Paper No. 3. World Bank, Washington D.C.

Mendoza, R., Bastiaensen, J., 2003. Fair trade and the coffee crisis in the Nicaraguan Segovias. Small Enterprise Development 14 (2), 36–47.

Murray, D. L., Raynolds, L. T., 2006. The future of fair trade coffee: Dilemmas facing Latin America's small-scale producers. Development in Practice 16 (2), 179–191.

Mutersbaugh, T., 2002. The number is the beast: A political economy of organic-coffee certification and producer unionism. Environment and Planning A 34, 1165–1184.

Neilson, J., 2008. Global private regulation and value-chain restructuring in Indonesian smallholder coffee systems. World Development 36 (9), 1607–1622.

Pelupessy, W., Díaz, R., 2008. Upgrading of lowland coffee in Central America. Agribusiness 24 (1), 119–140.

Philpott, S. M., Bichier, P., Rice, R., Greenberg, R., 2007. Field-testing ecological and economic benefits of coffee certification programs. Conservation Biology 21 (4), 975–985.

Ponte, S., 2002a. The 'latte revolution'? regulation, markets and consumption in the global coffee chain. World Development 30 (7), 1099–1122.

Ponte, S., 2002b. Standards, trade and equity: Lessons from the specialty coffee industry. CDR Working Paper 2 (13), 1–43, cited By (since 1996): 2 Export Date: 18 January 2008 Source: Scopus.

Ponte, S., Ewert, J., 2009. Which way is "up" in upgrading? Trajectories of change in the value chain for South African wine. World Development 37 (10), 1637–1650.

Ponte, S., Gibbon, P., 2005. Quality standards, conventions and the governance of global value chains. Economy and Society 34, 1–31.

Raynolds, L. T., 2000. Re-embedding global agriculture: The international organic and fair trade movements. Agriculture and Human Values 17 (3), 297–309.

Raynolds, L. T., 2002. Consumer/producer links in fair trade coffee networks. Sociologia Ruralis 42 (4), 404–424

Raynolds, L. T., Murray, D., Heller, A., 2007. Regulating sustainability in the coffee sector: A comparative analysis of third-party environmental and social certification initiatives. Agriculture and Human Values 24 (2), 147–163.

Raynolds, L. T., Murray, D., Taylor, P. L., 2004. Fair trade coffee: Building producer capacity via global networks. Journal of International Development 16 (8), 1109–1121.

Renard, M.-C., 2005. Quality certification, regulation and power in fair trade. Journal of Rural Studies 21 (4), 419–431.

Rigby, D., Cáceres, D., 2001. Organic farming and the sustainability of agricultural systems. Agricultural Systems 68 (1), 21–40.

Riisgaard, L., Bolwig, S., Ponte, S., Du Toit, A., Halberg, N., Matose, F., 2010. Integrating poverty and environmental concerns into value-chain analysis: A strategic framework and practical guide. Development Policy Review 28 (2), 195–216.

Slob, B., 2006. A fair share for coffee producers. In: Osterhaus, A. (Ed.), Business unusual. Successes and challenges of fair trade. FLO, IFAT, NEWS, and EFTA, Newcastle-upon-Tyne, pp. 121–139.

Talbot, J. M., 1997. Where does your coffee dollar go? The division of income. Studies in Comparative International Development 32 (1), 56–91.

Talbot, J. M., 2002. Tropical commodity chains, forward integration strategies and international inequality: Coffee, cocoa and tea. Review of International Political Economy 9 (4), 701–734.

Tallontire, A., 2009. Top heavy? Governance issues and policy decisions for the fair trade movement. Journal of International Development 21 (7), 1004–1014.

Wills, C., 2006. Fair trade: What's it all about? In: Osterhaus, A. (Ed.), Business unusual. Successes and challenges of fair trade. FLO, IFAT, NEWS, and EFTA, Newcastle-upon-Tyne, pp. 7–27.

Wollni, M., Zeller, M., 2007. Do farmers benefit from participating in specialty markets and cooperatives? The case of coffee marketing in Costa Rica. Agricultural Economics 37 (2-3), 243–248.

— Chapter 8 —

Summarizing Discussion and Conclusions

Coffee certification schemes, managed by farmer organizations, are supported by governments and international donors in order to link smallholder farmers to high-value markets, increase farmers' incomes, diminish power and information asymmetries, and to contribute to poverty reduction. Yet, there is only weak empirical evidence available which proves these desired effects of certification schemes. Apart from some exceptions, the available studies usually neither apply random sampling strategies of farmers or their organizations/cooperatives nor have they compared several cooperatives and certification standards simultaneously. This research contributed to filling the existing knowledge gaps through an analytical comparison of conventional, organic and Organic-Fairtrade certified small-scale coffee farmers, cooperatives, and value chains. The comparison has been based on two analytical levels: (i) on the smallholder household level and (ii) on the organizational and institutional level with regard of the cooperatives and respective coffee value chains. The study first identified the socio-economic costs and benefits of participation in organic and Organic-Fairtrade certified coffee chains with respect to coffee and household incomes as well as household poverty. Second, it was examined which role the farmer organizations, their respective business models and upgrading strategies, play for the success or failure of certification schemes. Third, the integration of coffee farmers and their cooperatives into the certified coffee value chain, the structure and functioning of the conventional and certified coffee value chains and the value adding effect of certification were explored.

8.1. Summary of major empirical results

The smallholder household level

In the research region, the coffee yields of conventional and certified coffee smallholders are usually 40% to 50% lower than national and regional average due to limited maintenance activities and inadequately managed coffee plantations. Highest yields (on average around 480kg/ha) are achieved by organic producers but yield levels vary, like for conventional and Organic-Fairtrade certified producers, between the cooperatives (ranging from 293kg/ha to 526kg/ha). In comparison to conventional prices, Organic-Fairtrade certified coffee achieved on average 11% and organic coffee 8% higher farm-gate prices. Certification schemes thus offer higher farm-gate prices, but price differences between cooperatives exist. Organic production processes require fewer purchased inputs but are more laborious. Due to constrained availability of family labor, additional labor has to be hired which offsets saved input costs. The higher prices of certified coffees compensate for production costs but fail to increase per hectare gross margins and profits in the case of Organic-Fairtrade farmers compared to conventional producers. Due to their higher yield levels, organic producers experience an increase in per hectare gross margins and profits. They have with 328US$/ha a significantly higher economic profit than Organic-Fairtrade farmers (147US$/ha) and conventional farmers (191US$/ha). Yet, as they tend to have smaller coffee areas and larger family sizes, the increase in gross margins does not result in improved per capita net coffee incomes for organic certified producers. Also Organic-Fairtrade certified producers do not have higher per capita net coffee incomes than conventional producers. Results of the qualitative interviews showed that at given yield levels, organic and Organic-Fairtrade farm-gate coffee prices are not sufficient to offset production costs in case coffee producers would apply an optimal organic farm management.

Among organic and Organic-Fairtrade certified producers, a higher share of households is grouped below the extreme poverty line than among conventional producers (45% compared to 30%) – which means that they cannot cover their food requirements. Between 60% and 70% of conventional and certified coffee producers are below the national poverty line. This indicates that the organic and Organic-Fairtrade coffee certification as such does not help northern Nicaraguan coffee farmers to earn a coffee income above the poverty line or to make them better off than their conventional fellow men. However, this result may also be driven by selection bias as coffee producers can choose whether they want to become certified or not. The self selection into coffee certification schemes may be based, for example, on the producers' asset endowment, which would be an observable selection variable, or on entrepreneurial skills and expected gains, which are unobservable selection variables. In the presented case, negative selection bias could have led to the higher poverty rates among the certified producers when, for example, the poorer or the less entrepreneurial skilled producers self-selected them into certified coffee schemes. Moreover, the study showed that over a period of ten years, organic certified producers became relatively poorer. In the year 1997, all groups had similar relative poverty levels. The Organic-Fairtrade certified producers first improved their relative poverty status during the coffee crisis (in 2002) and were relatively better off than conventional producers. Since then, the relative poverty levels of Organic-Fairtrade producers deteriorated compared to conventional producers. Yet again, the results may be influenced by negative selection bias due to self-selection into the certification schemes.

Irrespective of whether farmers were certified or not, the hardest time of the year starts two to three months after the last coffee harvest is sold. From May onwards, Nicaragua's coffee smallholders face two to three months of food shortages since by that time own stocks are depleted and food prices are high. As coffee farmers have no savings, they face liquidity constraints once the income from the coffee sales is spent. Diets shift from maize and beans to plantain since it is intercropped with the coffee. Due to beginning of the rainy season, farmers face more health problems while there are high labor requirements on their farms. Farmers use several strategies to cope with this situation. Some farmers, who only grow coffee and have small land areas, seek sporadically work at neighboring farms. A more typical strategy is to obtain a credit from a cooperative or microfinance institution. When credit needs are higher than the approved credit, farmers request credits from other microfinance institutions or informal money lenders. In many cases the credit is used for immediate consumption needs, like food or medicine, and only partially invested in the farm. Consequently, harvested yields stay low, leading to low incomes and new credit requirements. When farmers are financially illiterate or requested higher credits than their payment capacity, they are likely to enter a vicious cycle of indebtedness. Farmers mentioned the lack of access to long-term credits which are needed for replanting parts of their coffee area as younger and higher density plantations could contribute to increase yield levels.

The organizational and institutional level with regard of the cooperatives and value chain

Each cooperative has a unique business model; they differ, for example, in member size, functions and services, internal organization, and financial characteristics. Despite their different business models the cooperatives often choose the same upgrading strategies as other cooperatives. The cooperatives have certain strengths, weaknesses, opportunities, and threats (SWOTs) in common but there are also cooperative specific SWOTs. The latter needs to be addressed internally by the cooperative's management. The common strength of the cooperatives is the quality potential of the region. The common weaknesses relate to the lack of credit access, a weak extension system, and weak rural infrastructure. The common threats of the cooperatives are the strong competition between national coffee buyers and the cooperatives, corruption and mismanagement, and, according to the qualitative interviews, increasing microclimatic variations and unreliable rainfall patterns which can contribute to severe harvest losses. The common opportunities range from more horizontal coordination to the introduction of members' promotion plans, share certificates acknowledging the member's possessions in the cooperative and increased transparency about deductions on payments. The latter measures serve to enhance the members' satisfaction and commitment to their cooperative. There is no obvious association between the coffee certification strategy of a cooperative and the coffee gross margins of its members. The number and importance of strengths and weaknesses as well as the amount of coffee-related services offered to producers tend to be more related to gross margins than the organic or Organic-Fairtrade certification.

Farmers are found to have no bargaining power over prices irrespective of the value chain, while certified cooperatives have limited bargaining power towards their buyers compared to the cooperative conventional chain. Power is unequally distributed between buyers and sellers of coffee in all chains. The quantity and quality of information flows depends on the cooperative and value chain model. Information asymmetries are fewer in certified chains, yet also depend on the cooperative.

Certified value chains tend to have more and smaller-sized actors, especially in consuming countries, compared to the conventional chain. This increases transaction costs in the certified value chains and thus leads to substantially lower producers' share of the final coffee retail price (8%-15% in certified chains compared to 24%-34% in conventional chains). This result is contrary to the claims of Fairtrade advocacy. The producers' share of retail price is also influenced by the conventional coffee prices. Good conventional prices lead to higher shares of the final retail price for producers in conventional chains than when conventional coffee prices are low. Thus, the Fairtrade certification does not change the current distribution of value that characterizes the mainstream coffee market. The value added through certification can easily evaporate at the cost of the primary producer when certified value chains are characterized by more and smaller-sized chain actors in consuming countries and when information and power asymmetries as well as governance problems in intermediating cooperatives exist.

8.2. Comparison of research outcome with results of other studies

Since this research focuses on filling prevailing knowledge gaps only some results have been also investigated by other researchers. This section compares their research results with own results in the respective areas.

Many researchers observed higher farm-gate prices for certified coffee during or after the coffee crisis (Arnould et al., 2009; Bacon, 2005; Bacon et al., 2008; Barham et al., 2011; Becchetti and Costantino, 2008; Raynolds et al., 2004; Valkila, 2009; Wollni and Zeller, 2007). Yet, Kilian et al. (2006), Philpott et al. (2007), and Utting-Chamorro (2005) do not consistently locate premiums for certified coffee for this period. In line with the above mentioned studies the presented research found price premiums and thus higher farm-gate prices for certified coffees. The double certified Organic-Fairtrade producers received even higher premiums than organic certified producers. Yet, it was also shown that premiums varied between the cooperatives and, furthermore, that farmers usually did not sell all their coffee at these high prices. Thus, the average farm-gate coffee prices producers received across all marketing channels was lower than the maximal price paid by a cooperative. When producers were grouped according to certification, there was no difference in average farm-gate coffee prices between organic and Organic-Fairtrade certified producers, but both obtained higher prices than conventional producers. The comparison of cooperatives showed that members of one Organic-Fairtrade cooperative obtained nearly the same average farm-gate coffee price as conventional producers. This case shows the need for random sampling of several cooperatives and not only of farmers – an approach which most of the above mentioned studies lacked – exceptions are Wollni and Zeller (2007) and Arnould et al. (2009).

Many of the above mentioned studies deduced that the price premiums for certified coffee lead to higher farm income which then reduces poverty. As net coffee income is not only determined by the price but also by yield levels, the required coffee quality, and production costs, we showed that this conclusion can be wrong. The findings that organic farmers have either higher production costs than conventional farmers (Kilian et al., 2006; Van der Vossen, 2005) or lower production costs (Nigh, 1997; Valkila, 2009) are different to the presented results. This research demonstrates that coffee producers in northern Nicaragua have similar total variable production costs (ranging from 245US\$/ha to 308US\$/ha)when family labor is excluded. Yet, certified coffee producers used significantly more person-days per hectare coffee (140-156days/ha compared to 104days/ha of conventional producers)), in part this was covered by more family and hired labor. Mutersbaugh (2002) stated that organic coffee production is only successful when farmers have high yields; the presented results also point in this direction. Bray et al. (2002) indicated that higher prices for organic coffee offset higher production costs and that farmers benefit from participation. The significantly higher economic profit of 328US\$/ha of organic farmers in Nicaragua confirms Bray et al. (2002), yet, paradoxically, the economic profit for Organic-Fairtrade farmers rejects their results as their profit was 147US\$/ha compared to 191US\$/ha of conventional farmers. Given this heterogeneity within the organic production system it is premature to conclude here about the economic benefits of organic production for Nicaragua's coffee smallholders.

Bacon et al. (2008) identified continuing poverty among certified coffee smallholders in another research region in Nicaragua, which is in line with the presented quantitative measurements of household income and their absolute as well as relative poverty levels. Regarding the producers' share of final retail price, this research showed for conventional farmers shares between 18% and 34%, depending on the value chain and year. These shares are higher than the 10% to 20% usually indicated in literature, but are still within the range of past peaks (Daviron and Ponte, 2005; Fitter and Kaplinsky, 2001). In the Fairtrade chains, conventional producers had a share of the final retail price of 8% and organic producers of 9% to 15%. These shares are similar to those certified coffee producers had obtained before and during the coffee crisis but lower than the conventional shares when prices are good (Mendoza and Bastiaensen, 2003).

8.3. Reflections on research methods, processes and results

This section reflects on the choice of research methods, on the influence of the researcher on the research process and outcome as well as on the research results. A small section is added with the personal opinion of the author contemplating on the meaning of the research results for consumers as this has been a commonly raised question when research results were presented.

8.3.1. Reflection on the choice of research methods

The choice of the applied quantitative and qualitative research methods was determined on the one hand by the literature review which showed that such an approach has not yet been pursued and that either qualitative or quantitative approaches alone were insufficient to base policies on it. On the other hand, the choice was determined by previous research and farming experience of the author leading to the attitude that the agricultural sector and farmers' decision making are complex and require both approaches to fully reflect reality. The quantitative approach with random sampling was necessary to be able to generalize the findings for the research region but also to provide a substantiated state of knowledge on economic benefits, gross margins, household incomes and poverty. The structured household survey though does not record well non-monetary aspects like the reasons of farmers selling their coffee also to other market channels when the cooperative pays much higher prices or the farmers' experiences with coffee certification and the organic production system. The focus groups were chosen to complement the semi-structured individual interviews to obtain a broader range of opinions, experiences, and ideas and thus to triangulate information and to reduce the problem of sample selection bias in the qualitative interviews. Observational, narrative or visual images were considered to be insufficient to identify farmers' experiences with conventional and certified coffee farming and marketing. The inclusion of the qualitative part, especially the questions related to gender, may have been additionally influenced by the author's gender and interest in these topics.

8.3.2. The influence of the researcher on the research process and outcome

The Nicaraguan culture and language was different than the author's mother tongue and culture. This is connected with advantages and disadvantages. On the one hand it is more difficult to capture the subtle language differences and messages between the lines. Also cultural ways in representing information and the perception of its content vary, so that it is necessary to 'decode' information. The many years of experience in Latin-America and informal talks with Nicaraguans, Latin-Americans and foreigners working in Nicaragua contributed to a better understanding and reflection of this information. The recording of the interviews further helped in language issues as interviews could be listened to several times. Yet, the presence of the recorder also irritated some respondents at the beginning but they soon seemed to be comfortable with it. Only one respondent opted for the choice of not having the interview recorded. The fact that interviews were recorded might have influenced, especially at the beginning, the responses. For this reasons, the interviews were started with a 'warm-up phase' through some casual topics about coffee production, the weather, the village and so on until respondents felt comfortable. Being a foreigner, the farmers might have been more hesitant to share all their wealth of information than to a local because they did not know how far they can trust. On the other hand, the 'outsider' position of the researcher may have also contributed to obtain information the farmers would not tell a local because class differences were less obvious, a local is more connected to other networks and might be from another political party. The author was perceived as having a neutral position and many farmers stated that they felt honored to be visited. Many farmers wanted the author to be kind of a spokesman for their stories and asked for their stories to be told abroad. Some coffee producers stated that they even appreciated the detailed questions, about e.g. production costs and income, as it helped them to reflect about these issues and emphasized their importance.

As a foreign researcher the author obtained a wealth of information although having been warned by some local key-persons, with whom the questionnaires were discussed, that sometimes honest answers to certain questions could not be expected, for example the use of credits. Yet, answers of coffee producers reflected what key-persons, informal talks, and field observations showed. So while acknowledging that for any type of interviews there is the potential of respondents' to either overstate or understate some answers depending on their expectations, data are believed to be reliable.

Some of the interviews with female cooperative members were different to those with the male producers and it was more difficult to establish a level of trust, especially when the women did not have an own piece of land. Some women have been very shy in the interviews, and the time frame of one interview, of one visit, was not enough to overcome this barrier between them and the author. The author noticed that women and related gender issues in this research region require a different research approach than semi-structured interviews, probably based on more observational and narrative methods and repeated field visits to establish a relationship of trust.

8.3.3. Reflection on the research results

The quantitative research results were triangulated with the qualitative results, and vice versa. They were complemented by many field observations and informal farm visits of small and larger coffee plantations, of cupping events, coffee 'ferias', local coffee network meetings, and even a visit to a Nicaraguan coffee conference. There were no points where the author found major differences to the quantitative or qualitative data. The fact that quantitatively and qualitatively interviewed farmers were members of randomly selected cooperatives further contributes to the validity of results.

The large standard deviation in the data (for example see Chapter 4 or Chapter 6) made it sometimes difficult to identify significant differences among conventional and certified producers. Yet, these standard deviations seem typical for Central America (Kilian et al., 2006). A higher number of cooperatives in the sample design would have been desirable, yet in countries like Nicaragua not all cooperatives are registered. Thus it was impossible to establish a cooperative list for the whole country to randomly select from and be able to generalize findings at country level. Although our sample design does not allow for generalizations beyond the study area of northern Nicaragua the results are likely to be representative for coffee growers in higher altitudes in Nicaragua as production conditions are similar in these areas. Field trips to lower regions showed that results are not transferable as coffee quality and thus prices but also growing conditions change.

This study does not provide for a causal econometric impact analysis of certification effects. Yet, only a causal analysis could control for selection bias, test for the causality and thus statistically verify the claims that certified coffee production contributes to poverty reduction. Since such a causal analysis will deliver further insights into the effects of certification, the author has started to conduct such an analysis using the local instrument variable approach for polychotomous choice models to control for unobserved selection bias.

Visits to coffee producers in Mexico and Colombia during the field research suggested that effects of certification seem to vary between countries as the institutional setting and coffee productivity varies. The visits led to the assumption that organic and Fairtrade coffee certifications can have more socio-economic benefits for farmers than in Nicaragua when there is a better organizational and rural infrastructure and when yields levels and human capacities of farmers and cooperatives are higher. Yet, it has to be questioned if these farmers are still the 'marginalized poor' producers which a Fairtrade buying consumer imagines to support. The results for coffee can not be generalized to other organic or Fairtrade certified products as the production system (smallholders, large-scale plantations), relevant rules, standards, and certification costs vary.

The research analyzed organic and Organic-Fairtrade certification schemes in coffee production from the viewpoint of development policy research and aimed at providing policy recommendations for coffee certifications. It is not the task of this research to interpret results in order to provide recommendations for consumers. Yet, when presenting the results in conferences, workshops and seminars, the question whether consumers shall continue or stop buying certified (coffee) products was commonly raised. For organic products the question arises less frequently because consumers usually buy organic because of environmental, health and food safety considerations and much less because they want to make a contribution to the livelihoods of marginalized producers. The latter is what consumers of Fairtrade products aim for. Therefore, a few phrases will be added at this point which reflect the personal opinion of the researcher in that respect - it is only referred to small-scale coffee production. The promises of organic and Fairtrade coffee certifications to contribute to poverty reduction and improve the livelihoods of marginalized smallholder producers are considered to be set too high. Yet, this should be no reason not to buy this kind of coffee. There might be still economic benefits for producers; it may not be the poorest and most marginalized ones which benefit but probably those around or slightly above the poverty line. For a consumer, certified coffee is at least a way where she or he can 'vote with the trolley' and take a symbolic stand against the mainstream production and marketing conditions. The choice of consumers determines the coffee buyers' behaviors, and as the coffee value chains are buyer driven, the consumer has an influence on the production and marketing conditions of coffee. The success of organic and Fairtrade certification schemes has motivated large companies like 'Starbucks' and 'Nestlé' to start with sustainable coffee certification and has also contributed to the development of new certification schemes which aim at improving the bottlenecks of the organic and Fairtrade certification schemes. Some are now long enough in the market that it is possible to evaluate their impacts, advantages and disadvantages over the organic and Fairtrade schemes. Yet, these schemes are, like the organic and Fairtrade certification scheme, developed in industrialized countries and are thus likely to reflect more the ideas, conceptions and needs of the industrialized countries than those of the smallholders in developing countries. As (Paarlberg, 2009, p 182) criticizes: *"Foreign suppliers that want to sell to the rich must produce the goods the rich demand, at the quality level they demand, and even with the production processes they demand. This implies honouring the regulatory standards - both the product standards and production process standards - favoured by the rich countries"*. Especially for the Fairtrade certification, it is necessary to reflect on the whole structure of their value chains, to publicly discuss what is actually meant by 'fairness', which rules are influenced by ideology and which changes are nowadays necessary to make the Fairtrade coffee certification more beneficial for smallholders (for example a reduction of value chain actors in consuming countries). It is worthwhile to further discuss whether Fairtrade should open its coffee standard to large plantations like it has been done for banana or tea. This would benefit the landless day laborers, which are usually even poorer than smallholders, while it is likely to squeeze smallholders out of the certification schemes – which is the reason cooperatives are against such an opening. More research is necessary to model if coffee day laborers benefit more from plantation certification than smallholders do; first results indicate this direction (Maertens and Swinnen, 2009)[1].

To change existing structures requires large human and financial resources and, thus, needs the support of consumers. There is no clear-cut answer to the question of whether to buy organic or Fairtrade products. It depends on the personal attitudes why these products are bought, what the consumer aims at achieving and how much ideology is behind. As the research results for Nicaragua showed both certification schemes need to be improved in order to effectively contribute to improving the livelihoods of marginalized smallholder producers. Yet, the question remains if any production and certification scheme should and could compensate for structural failings at the economic, political, and social level in a given country.

---

[1]For example, research by Maertens and Swinnen (2009) showed that in Senegal poorer households benefited from agricultural export development following certain production standards rather through the labor than through the product market.

8.4. Conclusions

The profitability of the organic certified coffee production system is not clear cut; when farmers are grouped according to certification status, there is a trend that organic but not Organic-Fairtrade certified producers achieve higher gross margins and profits than conventional farmers in northern Nicaragua. The results depend strongly on each cooperative. This research shows that higher farm-gate prices do not necessarily lead to higher per capita net coffee income, as yield levels, production costs, family and land size, as well as labor availability play important roles. Therefore, higher farm-gate prices for certified coffees are not sufficient to increase per capita incomes. Coffee yields, profitability and efficiency need to be increased, as prices for certified coffee cannot compensate for low productivity, land or labor constraints. Long-term credits to invest in replanting old and low density coffee plots are one option to increase yields but risk at the same time to enforce the vicious cycle of indebtedness when the general farm management is inadequate and farmers do not earn sufficient to pay back interests and credits. Provision of cheap organic fertilizer by cooperatives can also contribute to increase yields.

The qualitative evaluation showed that the gross margins per hectare depend on the business model of a cooperative and thus the (coffee-related) services it offers, its respective SWOTs, and upgrading strategies of which the certification strategy is one. Organic or Organic-Fairtrade certification as an upgrading strategy seems only then successful when the business model of a cooperative, its strengths, weaknesses, and other upgrading strategies are supportive. Given the constraints mentioned above, a well functioning cooperative is a necessary but not sufficient condition as the example of one well run Organic-Fairtrade certified cooperative with low gross margins showed. The organizational structure of cooperatives, their human and financial capacities are important for coffee marketing. In the case of northern Nicaragua's producers, coffee upgrading through participation in certification schemes is not as beneficial regarding income and value distribution as assumed by governments, NGOs, and international donors. In certified coffee value chains the added value at retail level tends to evaporate at producers' level due to the many actors in consuming countries. The conventional coffee production and value chain should be reconsidered by researchers, policy-makers, and NGOs as a system which – with the right chain structure and institutional arrangements – could also benefit producers due to higher efficiencies and lower transaction costs.

The main causes of continuing poverty among smallholder coffee growers in Nicaragua seem not the lack of market access or so-called 'unfair' trading conditions. Based on the qualitative analysis, reasons for poverty are lack of entrepreneurial and management skills of farmers and cooperative staff, financial illiteracy and indebtedness of farmers as well as a very weak rural infrastructure. Based on the quantitative results potential reasons for poverty are low yield and productivity levels, land and labor constraints. The cooperatives' business models, management capacities and decisions as well as their upgrading strategies influence the coffee income of farmers but weak rural infrastructure with an insufficient road, transport, health, electricity, phone and schooling system further plays an important role. Certification schemes do not address or are able to solve these problems. Prices for certified coffee cannot compensate for low productivity, land or labor constraints. Therefore, certification schemes can only be part of a viable development policy for poor small-scale farmers in northern Nicaragua; the production, infrastructural, organizational and institutional problems mentioned above require even more attention from policy makers.

8.5. Outlook for further research and policy implications

Further research which addresses certification schemes should focus on several certification schemes at the same time and should especially concentrate the recently emerging certification schemes. A comparison of coffee producers in several countries with the same research approach could more specifically identify factors and conditions which determine the economic success of certification schemes and should ideally include the institutional setting at national level. Further impact assessments of certification schemes on socio-economic effects on smallholders should include panel data as well as the institutional context in which farmers operate and explicitly analyze the business models and upgrading strategies of the respective farmers' organizations. Research on coffee value chains should identify possibilities to reduce transaction costs and raise efficiencies, not only in producing but especially in coffee consuming countries. The vicious cycle of indebtedness of smallholder coffee producers which was discovered in this research merits further investigation which addresses in detail the access to credits, nominal and real interest rates of credits, income and expenditure patterns to identify possible policy intervention areas and ways to alleviate the financial constraints of farmers. Finally, agricultural diversification possibilities in the higher altitudes are scarce. Research could contribute to find appropriate varieties of other crops which can be grown in addition to coffee and also have a market value.

It is recommended that policies which aim at increasing smallholder coffee incomes through upgrading should focus, apart from production aspects, on the institutional context of smallholders and their cooperatives. Regarding coffee production, policies should address the low yield levels, for example through research investments for improved, stress-tolerant and locally adapted varieties to encounter the microclimatic variations. Coffee quality in the region should be further strengthened through adequate national strategies and facilitation of investments in this regard. It is recommended to establish an efficient extension system equipped with adequate financing as cooperatives do not have sufficient funds of their own to deliver these services. This could be also the support of the establishment of extension associations which operate regionally and are financed by their members' contribution. Extension itself should especially target the entrepreneurial skills of farmers.

In order to better link farmers to (high-value) markets and to increase their income, it is recommended to focus more on the structure and functioning of producer organizations and their respective value chains. Business and strategic advice to cooperatives is necessary, as cooperative leaders and staff are not fit for international markets, in which they have to act. The organizational and marketing skills of cooperative staff need to be improved which also may enhance their bargaining power with other value chain actors in the long-term. It is important to increase the efficiencies of cooperatives; this could also include the downgrading of some functions taken in the value chain like the processing and exporting of coffee. Possibilities to increase cooperation with exporters should be further explored as exporters can play a crucial role through their expertise, financial resources, and economies of scale. A national or agricultural bank which provides credits to cooperatives (at market interest rates and lending conditions) would reduce the reliance and dependence on exporters or international credit providers and could ease liquidity constraints of cooperatives. Policies could further oblige cooperatives to be more transparent in their invoices and to introduce share certificates acknowledging the member's possession in the cooperative. This enhances transparency and members' control of their cooperatives. An obligatory annual external auditing of cooperatives, like it exists in other countries, is considered to be important to reduce mismanagement of a cooperative. It will also increase the creditworthiness of cooperatives for banks.

While not having been discussed in detail, a closer look at successful policies in the coffee sector from other countries, like Colombia's coffee federation, reveals that institutional support at national level is of great importance. Competitive advantage of smallholders and their cooperatives is not built over night, but must be based on long-term strategic development of the whole national coffee sector. With a growing but still limited demand in consuming countries, coffee certifications alone cannot help to improve producers' welfare in the long run because entry barriers are low and new producers will continue to enter the market until the profit margin is zero. Therefore, a supportive sector strategy needs to be effectively implemented at national level, which could include a coffee institute or federation like in Colombia or Costa Rica, and should be accompanied by investments in rural infrastructure, such as electricity grids, roads, internet and phone connections.

In order to further increase the producers' prices and improve the distribution of value shares, additional attention should be given to enhancing trade, processing, and marketing efficiencies in the organic and Fairtrade value chains in consuming countries. This could be a reorganization of the alternative trade sector with its many small profit or non-profit enterprizes and organizations processing and marketing organic and Fairtrade coffee. These actors could consolidate to exert economies of scale and thus reduce the transaction costs occurring in consuming countries. While this is certainly a new way of thinking in the alternative trade sector, streamlining could effectively contribute to its mandate of improving producers' shares of retail prices, raising prices and thus contributing to a more equitable benefit sharing.

— Chapter 9 —

Bibliography

Arnould, E. J., Plastina, A., Ball, D., 2009. Does fair trade deliver on its core value proposition? Effects on income, educational attainment, and health in three countries. Journal of Public Policy & Marketing 28 (2), 186–201.

Asfaw, S., Mithöfer, D., Waibel, H., 2010. Agrifood supply chain, private-sector standards, and farmers' health: Evidence from Kenya. Agricultural Economics 41 (3-4), 251–263, 1574-0862.

Bacon, C., 2005. Confronting the coffee crisis: Can fair trade, organic and specialty coffees reduce small-scale farmer vulnerability in northern Nicaragua? World Development 33 (3), 497–511.

Bacon, C., Ernesto Mendez, V., Gomez, M. E. F., Stuart, D., Flores, S. R. D., 2008. Are sustainable coffee certifications enough to secure farmer livelihoods? The Millennium Development Goals and Nicaragua's fair trade cooperatives. Globalizations 5 (2), 259–274.

Barham, B. L., Callenes, M., Gitter, S., Lewis, J., Weber, J., 2011. Fair trade/organic coffee, rural livelihoods, and the "agrarian question": Southern Mexican coffee families in transition. World Development 39 (1), 134–145.

Basu, A. K., Chau, N. H., Grote, U., 2003. Eco-labeling and stages of development. Review of Development Economics 7 (2), 228–247.

Becchetti, L., Costantino, M., 2008. The effects of fair trade on affiliated producers: An impact analysis on Kenyan farmers. World Development 36 (5), 823–842.

Bolwig, S., Gibbon, P., Jones, S., 2009. The economics of smallholder organic contract farming in tropical Africa. World Development 37 (6), 1094–1104.

Bray, D. B., Sánchez, J. L. P., Murphy, E. C., 2002. Social dimensions of organic coffee production in Mexico: Lessons for eco-labeling initiatives. Society & Natural Resources 15 (5), 429–446.

Cashin, P., McDermott, C. J., Scott, A., 2002. Booms and slumps in world commodity prices. Journal of Development Economics 69 (1), 277–296.

Codron, J.-M., Siriex, L., Reardon, T., 2006. Social and environmental attributes of food products in an emerging mass market: Challenges of signaling and consumer perception, with European illustrations. Agriculture and Human Values 23 (3), 283–297.

Daviron, B., Ponte, S., 2005. The coffee paradox. Global markets, commodity trade and the elusive promise of development. Zed Books, London and New York.

Fairtrade Foundation, 2011. The Arabica coffee market 1989-2010: Comparison of fairtrade and New York prices. Fairtrade Foundation, London, accessed 17 January 2011.
URL http://www.fairtrade.org.uk/includes/documents/cm_docs/2011/a/arabica_pricechart_89_jan11.pdf.

Fitter, R., Kaplinsky, R., 2001. Who gains from product rents as the coffee market becomes more differentiated? A value-chain analysis. IDS Bulletin 32 (3), 69–82.

FLO, 2007. Annual report 2007. An inspiration for change. Fair Trade Labelling Organizations International (FLO), Bonn, accessed 15 December 2010.
URL http://www.fairtrade.at/pics/infomaterial/FLO_AR2007.pdf.

FLO, 2008. Fairtrade standards for coffee for small farmers' organizations. Fair Trade Labelling Organizations International (FLO), Bonn.

FLO, 2010. Coffee: Market overview. Fairtrade Labelling Organizations International (FLO), Bonn, accessed 15 January 2011.
URL http://www.fairtrade.net/fileadmin/user_upload/content/2009/standards/documents/Coffee_MarketOverview_final_161210.pdf.

Garibay, S. V., Zamora, E., 2003. Producción orgánica en Nicaragua: Limitaciones y potencialidades. Servicio de Información Mesoamericano sobre Agricultura Sostenible (SIMAS), Managua, accessed 22 January 2011.
URL http://orgprints.org/2691/.

Gereffi, G., Humphrey, J., Sturgeon, T., 2005. The governance of global value chains. Review of International Political Economy 12 (1), 78–104.

Giovannucci, D., Pierrot, J., Kasterine, A., 2010. Trends in the trade of certified coffees. MPRA Paper No. 27551. International Trade Center, accessed 30 January 2011.
URL http://mpra.ub.uni-muenchen.de/27551/.

Giovannucci, D., Villalobos, A., 2007. The state of organic coffee: 2007 US update. Sustainable Markets Intelligence Center (CIMS), San Jose, accessed 15 December 2010.
URL http://orgprints.org/12856/1/State_of_Organic_Coffee_2007_US_update.pdf

Gobierno de Nicaragua, 2001. Mapa de pobreza extrema de Nicaragua 2001. Managua, accessed 18 Januar 2011.
URL http://www.inide.gob.ni/Pobreza/publicacion/mapapobreza2001.pdf.

González, A. A., Nigh, R., 2005. Smallholder participation and certification of organic farm products in Mexico. Journal of Rural Studies 21 (4).

Gresser, C., Tickel, S., 2002. Mugged: Poverty in your cup. Oxfam International; Make Trade Fair, Washington D.C., accessed 21 November 2006.
URL http://www.oxfamamerica.org/newsandpublications/publications/research_reports/mugged.

Henson, S., Reardon, T., 2005. Private agri-food standards: Implications for food policy and the agri-food system. Food Policy 30 (3), 241–253.

ICO, 2009. Organic coffee export statistics. WP Statistics 140/09. International Coffee Organization (ICO), London, accessed 22 January 2011.
URL http://www.ico.org/documents/wp-statistics-140e-organic.pdf.

IFOAM, 2006a. Organic agriculture and food security. International Federation of Organic Agriculture Movements (IFOAM), Bonn, accessed 20 July 2006.
URL http://www.ifoam.org/organic_facts/food/pdfs/Food_Security_Leaflet.pdf.

IFOAM, 2006b. Organic agriculture and rural development. International Federation of Organic Agriculture Movements (IFOAM), Bonn, accessed 15 December 2010.
URL http://www.ifoam.org/growing_organic/3_advocacy_lobbying/eng_leaflet_PDF/Rural_Development.pdf.

IFOAM, 2009a. Definition of organic agriculture. International Federation of Organic Agriculture Movements (IFOAM), Bonn, accessed 29 March 2010.
URL http://www.ifoam.org/growing_organic/definitions/doa/index.html.

IFOAM, 2009b. The principles of organic agriculture. International Federation of Organic Agriculture Movements (IFOAM), Bonn, accessed 15 December 2010.
URL http://www.ifoam.org/about_ifoam/principles/index.html.

IICA, 2004. Cadena agroindustrial del café en Nicaragua. Instituto Interamericano de Cooperación para la Agricultura (IICA), Managua.

Kilian, B., Jones, C., Pratt, L., Villalobos, A., 2006. Is sustainable agriculture a viable strategy to improve farm income in Central America? A case study on coffee. Journal of Business Research 59 (3), 322–330.

Lewin, B., Giovannucci, D., Varangis, P., 2004. Coffee markets: New paradigms in global supply and demand. Agriculture and Rural Development Discussion Paper No. 3. World Bank, Washington D.C.

Linton, A., 2008. A niche for sustainability? Fair labor and environmentally sound practices in the specialty coffee industry. Globalizations 5 (2), 231–245.

Maertens, M., Swinnen, J. F. M., 2009. Trade, standards, and poverty: Evidence from Senegal. World Development 37 (1), 161–178.

Mendoza, R., Bastiaensen, J., 2003. Fair trade and the coffee crisis in the Nicaraguan Segovias. Small Enterprise Development 14 (2), 36–47.

Murray, D. L., Raynolds, L. T., 2006. The future of fair trade coffee: Dilemmas facing Latin America's small-scale producers. Development in Practice 16 (2), 179–191.

Mutersbaugh, T., 2002. The number is the beast: A political economy of organic-coffee certification and producer unionism. Environment and Planning A 34, 1165–1184.

Mutersbaugh, T., 2005. Fighting standards with standards: Harmonization, rents, and social accountability in certified agrofood networks. Environment and Planning A 37 (11), 2033–2051.

Nigh, R., 1997. Organic agriculture and globalization: A Maya associative corporation in chiapas, mexico. Human Organization 56 (4), 427–436.

Paarlberg, R., 2009. Starved for science. How biotechnology is being kept out of Africa. Harvard University Press, Cambridge, Massachusetts, London.

Philpott, S. M., Bichier, P., Rice, R., Greenberg, R., 2007. Field-testing ecological and economic benefits of coffee certification programs. Conservation Biology 21 (4), 975–985.

Ponte, S., 2002. The 'latte revolution'? regulation, markets and consumption in the global coffee chain. World Development 30 (7), 1099–1122.

Ponte, S., Gibbon, P., 2005. Quality standards, conventions and the governance of global value chains. Economy and Society 34, 1–31.

Raynolds, L. T., 2000. Re-embedding global agriculture: The international organic and fair trade movements. Agriculture and Human Values 17 (3), 297–309.

Raynolds, L. T., 2009. Mainstreaming fair trade coffee: From partnership to traceability. World Development 37 (6), 1083–1093.

Raynolds, L. T., Murray, D., Taylor, P. L., 2004. Fair trade coffee: Building producer capacity via global networks. Journal of International Development 16 (8), 1109–1121.

Rice, R. A., 2001. Noble goals and challenging terrain: Organic and fair trade coffee movements in the global marketplace. Journal of Agricultural and Environmental Ethics 14 (1), 39–66.

Rigby, D., Cáceres, D., 2001. Organic farming and the sustainability of agricultural systems. Agricultural Systems 68 (1), 21–40.

Scoones, I., 1998. Sustainable rural livelihoods-framework for analysis. IDS Working Paper No. 72. Institute of Development Studies (IDS), Brighton.

Slob, B., 2006. A fair share for coffee producers. In: Osterhaus, A. (Ed.), Business unusual. Successes and challenges of fair trade. FLO, IFAT, NEWS, and EFTA, Newcastle-upon-Tyne, pp. 121–139.

Swinnen, J. F. M., Maertens, M., 2007. Globalization, privatization, and vertical coordination in food value chains in developing and transition countries. Agricultural Economics 37, 89–102.

Talbot, J. M., 1997. Where does your coffee dollar go? The division of income. Studies in Comparative International Development 32 (1), 56–91.

Talbot, J. M., 2002. Tropical commodity chains, forward integration strategies and international inequality: Coffee, cocoa and tea. Review of International Political Economy 9 (4), 701–734.

Tallontire, A., 2009. Top heavy? Governance issues and policy decisions for the fair trade movement. Journal of International Development 21 (7), 1004–1014.

Taylor, P. L., Murray, D. L., Raynolds, L. T., 2005. Keeping trade fair: Governance challenges in the fair trade coffee initiative. Sustainable Development 13 (3), 199–208.

Utting-Chamorro, K., 2005. Does fair trade make a difference? The case of small coffee producers in Nicaragua. Development in Practice 15 (3-4), 584–599.

Valkila, J., 2009. Fair trade organic coffee production in Nicaragua – sustainable development or a poverty trap? Ecological Economics 68 (12), 3018–3025.

Van der Vossen, H. A. M., 2005. A critical analysis of the agronomic and economic sustainability of organic coffee production. Experimental Agriculture 41 (4), 449–473.

Varangis, P., Siegel, P., Giovannucci, D., Lewin, B., 2003. Dealing with the coffee crisis in Central America. Impacts and strategies. Policy Research Working Paper No. 2993. World Bank, Washington D.C.

Varughese, G., Ostrom, E., 2001. The contested role of heterogeneity in collective action: Some evidence from community forestry in Nepal. World Development 29 (5), 747–765.

WFTO, 2010. Charter of Fair Trade Principles. Introduction. World Fair Trade Organization WFTO, Culemborg, accessed 22 January 2011.
URL `http://www.wfto.com/index.php?option=com_content&task=view&id=1082&Itemid=11`.

Willer, H., Yussefi, M. (Eds.), 2007. The world of organic agriculture. Statistics and emerging trends 2007, 9th Edition. International Federation of Organic Agriculture Movements (IFOAM) and Forschungsinstitut fuer biologischen Landbau (FiBL), Bonn.

Wills, C., 2006. Fair trade: What's it all about? In: Osterhaus, A. (Ed.), Business unusual. Successes and challenges of fair trade. FLO, IFAT, NEWS, and EFTA, Newcastle-upon-Tyne, pp. 7–27.

Wintgens, J. N. (Ed.), 2009. Coffee: Growing, processing, sustainable production. A guidebook to growers, processors, traders and researchers, 2nd Edition. Wiley-VCH, Weinheim.

Wollni, M., Lee, D. R., Thies, J. E., 2010. Conservation agriculture, organic marketing, and collective action in the honduran hillsides. Agricultural Economics 41 (3-4), 373–384.

Wollni, M., Zeller, M., 2007. Do farmers benefit from participating in specialty markets and cooperatives? The case of coffee marketing in Costa Rica. Agricultural Economics 37 (2-3), 243–248.

World Bank, 2008. Nicaragua poverty assessment. Volume 1: Main Report. No. 39736-NI. World Bank, Washington D.C.

Appendices

— Appendix A —

Household Questionnaire

Análisis de participación en las cadenas de comercialización del café certificado y sus efectos en el bienestar de los pequeños agricultores

Encuesta

de Tina Beuchelt (2007)

Contenido

<u>**PART 1: IDENTIFICACIÓN Y DATOS GENERALES**</u>

1.1	**Nombre de la Entrevistadora**	
1.2	**Fecha**	......./......./2007
1.3	**Número de Identification del hógar** *Por favor, anota en todos páginas!*	
1.4	**Departamento** *(Código 1)*	
1.4	**Municipio** *(Código 2)*	
1.5	**Nombre de la comunidad**	
1.6	**Nombre de la Cooperativa** *(Código 3) y, si aplica, nombre de la cooperativa de base)*	
1.7	**Nombre y apellidos del / de la jefa/ jefe del hogar**	
1.8	**Nombre y apellidos del / de la entrevistada/o**	
1.9	**Nombre de la finca**	
1.10	**Número del teléfono**	
1.11	**Número del celular**	

Datos georeferenciados

1.12	**Altura (metros)**	
1.13	**Coordinado del norte (latitude)**	
1.14	**Coordinado del oeste (longitude)**	

Código 1	*Código 2*	*Código 3*
Lista de Departmentos:	**Lista de Municipios:**	**Lista de Cooperativas:**
Nuevas Segovias........... (1) Madríz........................... (2)	San Juan de Río Coco (1.1) Telpaneca/Santo Dom.(1.2) Jícaro ....(2.1) Quilalí ....(2.2)	Caja Rural SJRC .... (1) Coop 20 de Abril............. (2) Coop Santiago............... (3) Coop Santo Domingo .... (4) Prococer .... (5) UCA SJRC..................... (6) UCPCO.......................... (7)

<u>Nota para la entrevistadora</u>: El hogar consiste de todas personas que viven abajo del mismo techo, comen de la misma olla y comparten los gastos. Es necesario que vivían juntos por lo menos 9 meses de los 12 meses pasados o que mandan remesas al hogar regularmente o que son hijos de la jefa/e y estudián en una escuela.

1

PART 2: MIEMBROS DEL HOGAR Y FUERZA LABORAL
2.1 Datos del Hogar / Demografía y Educación

ID	2.1.1 Nombres y apellidos	2.1.2 Sexo (1) Varón (2) Mujer	2.1.3 Año de nacimiento Año	2.1.4 Parentesco con el/la jefe/a de la familia (Código 4)	2.1.5 Estado Civil (Código 5)	2.1.6 Puede "…" escribir y leer? (Código 6)	2.1.7 Máximo nivel educativo que tiene "…"? (Código 7)	2.1.8 Si niños tienen entre 6-17 años "…" todavía va a la escuela? (1) Si (0) No	2.1.9 Si sí, va regularmente a la escuela? (Código 8)	2.1.10 Si no va regularmente a la escuela, porque no? (1) Falta de dinero (2) Tiene que trabajar (3) Demasiado joven (4) Mal estado de salud (5) Otro:______
1										
2										
3										
4										
5										
6										
7										
8										
9										
10										
11										

Código 4 Parentesco con el/la jefe/a
(1) Jefe/a
(2) Esposa/o
(3) Hija/o
(4) Madre o padre
(5) Abuela/o
(6) Nietos
(7) Nuera/yerno
(8) Hermana/o
(9) Otro familiar
(10) Otro no-familiar

Código 5 Estado civil
(1) Soltera/o
(2) Casado con esposo/a permamente viviendo en el hógar
(3) Casado con esposo/a migrante
(4) Acompañado/a
(5) Viuda/o
(6) Divorciada/o

Código 6 Escribir/leer
(1) No puede escribir ni leer
(2) Solo puede leer
(3) Puede escribir y leer

Código 7 Nivel educativo
(0) Nunca atendió escuela
(1) Primaria incompleta
(2) Primaria completa
(3) Secundaria incompleta
(4) Secundaria completa
(5) Técnico
(6) Universitario

Código 8 Atendencia escolar
(1) Todos los días/sabatino
(2) No regularamente
(3) Fue el año pasado pero no este

Part 2.2: EMPLEO Y INGRESOS

No.	2.2.1 Cuál es el la actividad principal de "…"? (CÓDIGO 9) Nombre Trabajo	2.2.2 Cúantas veces trabaja en la finca familiar? (1) Tiempo completo en propia finca (2) Tiempo temporal en propia finca y temporal en otro lugar (3) Ayuda en propia finca (4) Nunoa	2.2.3 Si el trabajo principal no es en propia finca/casa, cuánto fue el ingreso del trabajo principal desde Marzo 2006 hasta Feb 2007? *Ingreso o salarío por día/catorcena/ en total* *(por favor indica cantidad)*	2.2.4 Tenía "…" una segunda actividad desde Marzo 2006 hasta Feb 2007? (0) No →2.2.6 Sí → (CÓDIGO 9)	2.2.5 Cuánto es la contribución al ingreso de la familia de la segunda actividad? *Ingreso o salarío por día/catorcena/ en total* *(por favor indica cantidad)*	2.2.6 Recibe comida en su trabajo en la primero y/o segunda actividad? No ……. (0) Sí…….. (1) Actividad 1ero 2ndo	2.2.7 Cuántas comidas por día recibió? Número de comidas 1er activi dad / 2ndo activi dad
01							
02							
03							
04							
05							
06							
07							
08							
09							
10							
11							

CÓDIGO 9 Principal y segunda actividad

(1) Empleo autónomo en finca
(2) Empleo autónomo que no es en su finca
(3) Estudiante
(4) Empleado/a del gobierno
(5) Asalario en el sector agropecuario
(6) Asalario en el sector non-agropecuario
(7) Trabajo a jornal en el sector agropecuario
(8) Trabajo a jornal en el sector non-agropecuario

(9) Trabajo doméstico en propia casa (ama de casa)
(10) Empleado/a Domestico/a (en otra casa)
(11) Ejercito/policía
(12) Sin empleo, pero buscando empleo
(13) No quiere trabajar o jubilado/ reteriado
(14) Incapacidad, no capaz de trabajar

Trabajo doméstico: Incluye todos los trabajos dentro o alrededor de la casa con el objetivo de cuidar todos los miembros de la casa. Esto incluye actividades como la preparación de comidas, limpiar, lavar, cuidar niños, gente mayor de 60 años , enfermo o inválido en el hógar.

3

Part 2.3: ENFERMEDADES, GASTOS y MIEMBROS DEL HOGAR AUSENTE/MIGRANTE

No.	2.3.1 Cuántas días en los últimos 12 meses fue "…" enfermo/a y así incapaz de trabajar?		2.3.2!!!!!!!!!!!!!!!!!! Cúales eran los gastos para ropa y zapatos de cada persona en los últimos 12 meses? *Solo para miembros permanente del hogar (incluye costos de material)* C$	2.3.3 La familia recibió dinero/ remesas de un familiar o otra persona ausente en los últimos 12 meses? No ……. (0) Sí………. (1)	2.3.4 Si la familia recibe remesas, de quién los recibe? *Por favor, anota nombre y parentesco de esta persona con el jefe/a de la casa*	2.3.5 Cuánto dinero/remesas recibe la familia cada mes? U$
	Nombre	Días				
01						
02						
03						
04						
05						
06						
07						
08						
09						
10						
11						

4

PART 3: ASUNTOS DEL HOGAR

3.1 La casa: *También pregunta la misma pregunta por la situación hace 5 y 10 años.*

Núm ero		3.1.X.1 Actual	3.1.X.2 Como fue hace 5 años (2002)	3.1.X.3 Como fue hace 10 años (1997)
3.1.1	Un miembro de este hogar es el dueño de este casa? (0) No (1) Sí ➔3.1.5			
3.1.2	Está alquilando este casa por bienes, servicios, or dinero? (0) No (gratis) ➔ 3.1.5 (1) Sí			
3.1.3	Cuantó al año tiene que pagar en efectivo para alquilar este casa? ***Entrevistadora:*** *Si no paga en efectivo, anota "0".*	______C$	______C$	______C$
3.1.4	Si paga el alquiler con bienes y servicios, cuál es el valor de estos cada año? *Entrevistadora: Si no paga en efectivo, anota "0".*	______C$	______C$	______C$
3.1.5	Cuantas habitaciones tiene la casa? *(Incluye todos cuartos, sala, comedor – también si son estructuras diferentes).* **No incluye cocina**, *baños/ inodoro adentro de casa, letrina.*	Número		
3.1.6	De qué material es el techo? (1) Tejas de madera (2) Plástico (3) Tejas de barro (4) Zinc			
3.1.7	De qué material son las paredes? (1) Plastico (2) Enrejada/Empalencada (ramas y adobe) (3) Embarillada/embarada con tierra (4) Entablada (5) Adobe (6) Adobe cubierto de cemento (7) Bloque (8) Ladrillos (9) Bloque/ladrillos cubierto de cemento			
3.1.8	De qué material es el piso? (1) Tierra (2) Embaldozado (cemento) (3) Cemento enladrillado/ladrillo artificial			
3.1.9	Qué tipo de material usa principalmente para cocinar? (1) Hojas, paja, ramas, cascara de café (2) Leña que se colleciona (3) Leña que se compra (4) Gas (5) Electricidad			

3.1.10	Con qué se alumbra la sala en la noche? (1) No puedo financiar ningún tipo de luz en la noche (2) Ocote (3) Candiles (4) Candela (5) Lámpara de gas (6) Linterna con baterías pequeñas (7) Batería de vehículo (8) Electricidad (publico, coneción privada) (9) Panel solar (10) Planta electrica					
3.1.11	Qué tipo de letrina tiene? (1) Ninguna, van al campo, arbustos... (2) Letrina privada (3) Inodoro privado					
3.1.12	Dónde cocina normalmente las comidas? (0) Fuera de la casa (sin paredes y techo) (1) En una de las salas/habitaciones de la casa (2) En una cocina separada					
3.1.13	Tiene uno de los siguientes cosas en su casa? (0) No (1) Sí, coneción privada (2) Sí, coneción pública	**a.**Agua de tubería **b.**Electricidad **c.**Teléfono				

3.1.14	Cuánto pagó la familia al mes para los siguentes servicios?	*Entrevistadora: Si no paga nada, entra zero "0".*		
	a. Electricidad	_______C$	_______C$	No llenar
	b. Agua	_______C$	_______C$	No llenar
	c. Teléfono	_______C$	_______C$	No llenar

6

155

3.2. Consumo de alimentos
Sera recomendable si los dos, el jefe/la jefa y la esposa/el esposo, pueden responder estas preguntas juntos

3.2.1	En los últimos siete días, había un evento especial (por ejemplo un evento e la familia, gente invitado, fiesta), dónde servía una comida mejor/especial?	_______**fecha**
3.2.2	**Si no había una fiesta:** Cuántas veces al día se servió comidas a los miembros de la casa ayer y anteayer? **Si había una fiesta, la pregunta debe estar relacionada a los dos días antes de la fiesta**	# comida
3.2.4	**Si no había una fiesta en los <u>últimos siete días</u>:** En cuántas comidas principales se sirvió uno de los siguientes alimentos en los últimos siete días? **Si había una fiesta, la pregunta debe estar relacionada a los 7 días ANTES de la fiesta/ evento especial** a) Pollo / b) Carne de res / c) Carne de cerdo / d) Queso	a) b) c) d)
3.2.5	**Si no había una fiesta en los <u>últimos siete días</u>:** En los últimos siete días, cuántas veces una comida principal consistía solamente de tortillas o guineas *(sin arroz y frijoles)*?	# comida
3.2.6	En los últimos 30 días, había siempre suficiente comida para los miembros de la casa? Si dice no, cuantos días no había suficiente comida?	(1) Sí (0) No ➜ ***anota números de días.***
3.2.7	En general, con que frecuencia compra los sigiuentes alimentos? a) arroz b) frijoles c) maíz? (1) Diario / (2) Dos veces en la semana / (3) Semanal / (4) Quincenal/catorcenal / (5) Mensual / (6) Menos frecuente que mensual / (7) Consume de propio cosecha	a) b) c)
3.2.8	Cuántas libras de arroz, frijoles y maíz tiene actualmente en su casa para autoconsumo? a) Arroz b) Frijoles c) Maíz *Por favor, indica cantidad en libra*	a) b) c)
3.2.9	Cuántas semanas alcanza esta cantidad de arroz, frijoles y maíz? a) Arroz b) Frijoles *Por favor, indica cantidad en días o semanas (o cuantas libras consumen al día)* c) Maíz	a) b) c)
3.2.10	Un miembro o más del hogar no comió algunas de las comidas durante los últimos <u>12 meses</u> por falta de dinero? (0) No ➜ 3.3.1 .... (1) Sí	
3.2.11	Cuántas veces ocurrió esto en los últimos 12 meses? (1) Más que 180 días (2) Menos que 180 días pero más que 30 días (3) Menos que 30 días pero más que 10 días (4) Menos que 10 días	

7

3.3 RECURSOS
3.3.1 Recursos generales

	3.3.1.1 Cuánto tenía hace 10 años (1997)	3.3.1.2 Cuánto tenía hace 5 años (2002)	3.3.1.3 Cuánto tiene actualmente?	3.3.1.4 Valor si lo vendría ahora? C$
a. Vehículo				
b. Motocicleta				
c. Bicicleta				
d. Carreta (de bueyes)/coche				
Electrodomésticos				
e. Televisión				
f. DVD /Videograbadora (VHS)				
g. Radio				
h. Telefóno (linea fija)				
i. Celular	No llenar	No llenar		
j. Refrigerador				
k. Ventiladores/albanico				
l. Horno de gas o electricidad				

3.3.2 Recursos de la finca

	3.3.2.1 Cuánto tenía hace 10 años (1997)	3.3.2.2 Cuánto tenía hace 5 años (2002)	3.3.2.3 Cuánto tiene actualmente?	3.3.2.4 Valor si lo vendría ahora? C$
Medios de Producción de la Finca				
a. Arado manual				
b. Arado mecanizado				
c. Almacén/Bodega				
d. Despulpadora manual				
e. Despulpadora eléctrico/con motor				
f. Pilas de fermentación				No llenar
g. Bomba de mochila				
Animales				
h. Ganado (Vaca, buey)				
i. Cerdo				
j. Cabra/Oveja/Peliboy				
k. Caballo/burro				
l. Gallinas				
m. Otro_____________				

8

3.3.2.5	Desde Marzo 2006 hasta Febrero 2007, vendío animales?	(0) No (1) Sí
3.3.2.6	Si vendió, cuánto fue su ingreso obtenido de la venta de estos animales?	__________C\$ en total
3.3.2.7	En el mismo tiempo, vendío productos de animales como cuajada, leche, huevos?	(0) No (1) Sí
3.3.2.8	Si vendió algo, cuánto fue el ingreso total de esta venta desde Marzo 2006 hasta Febrero 2007?	__________C\$ en total

3.3.3 Tenencia de la tierra

3.3.3.1 Estába arrendando terreno desde Marzo 2006 hasta Febrero 2007?   (0) No   (1) Sí

		3.3.3.1.X.1 Cuántas manzanas (mz)	3.3.3.1.X.2 Precio (C\$ por mz o total)
3.3.3.1.1	Area arrendado para producción personal		
3.3.3.1.2	Area arrendado/prestado a otra persona		

3.3.3.2 En los últimos 5 años, compro o vendío terreno?   (0) No   (1) Sí

		3.3.3.2.X.1 Cuántas manzanas? (mz)	3.3.3.2.X.2 En que año?	3.3.3.2.X.3 Precio (C\$ por mz o total)
3.3.3.2.1	Terreno comprado			
3.3.3.2.2	Terreno vendido			

3.3.3.3	Cuántas fincas poseen los miembros del hogar?	
3.3.3.4	Tiene documentos legales de su finca? (0) Ninguno (1) Escritura pública *(un título registrado y catastrado)* (2) Escritura privada (3) Título de reforma agraria	
3.3.3.5	Si posee documentos legales, en que año le recibió/compró?	_______año
3.3.3.6	El documento está registrado en el nombre de quién? (1) Jefa/e de la casa (2) Esposa/o (3) Jefa/e de la casa y esposa/o	
3.3.3.7	Si no tiene un documento individual, su cooperativa tiene un título registrado?	(0) No (1) Sí (2) No sabe
3.3.3.8	También tiene tierra sin tener documentos legales?	(0) No (1) Sí
3.3.3.9	Si tiene tierra sin documentos legales, qué tamaño tiene esta area? Area (mz)	__________mz

9

		3.3.3.X.1 Area hace 10 años (1997) (mz)	3.3.3.X.2 Area hace 5 años (2002) (mz)	3.3.3.X.3 Area (actual) (mz)	3.3.3.X.4 Valor si lo vendría ahora? en C$/ area total o mz
3.3.3.10	Cuántas mz tiene su **finca** en total? *(incluye todo, bosque, poteros..* ***AREA DE TODOS FINCAS)***				___________CS$
3.3.3.11	Si no vive en su finca, cuántas mz tiene el terreno **de su casa**?				___________CS$

3.3.3.12	A que distancia está su finca de la casa?	(0) Casa está en finca ________Distancia en km (o ________horas________minutos **caminando**)
3.3.3.13	A que distancia está su casa del pueblo más cercano?	(0) Casa es en la comunidad (pueblo) ________Distancia en km (o ________horas________minutos **caminando**)
3.3.3.14	A que distancia está su casa del servicio médico más cercano (sea centro de salud o hospital)?	________Distancia en km (o ________horas________minutos **caminando**)
3.3.3.15	A que distancia está su casa del mercado/pulpería más cercano?	________Distancia en km (o ________horas________minutos **caminando**)

10

PART 4: PRODUCCIÓN AGRICOLA

4.1 Fuentes de ingreso

4.1.1	Con cúal de estos rubros se mantiene usted y su familia? (1) Venta de café (2) Producción y venta de granos básicos/hortalizas (3) Venta de animales o de sus productos (4) Remesas/dinero enviado de familiares migrantes (5) Venta de madera/leña (6) Venta de fuerza de trabajo (7) Otro (especifica)______	Prinipal	Otros
4.1.2	Cuánto fue el ingroo de la venta de la leña desde Marzo 2006 hasta Febrero 2007?	______ C$ en total	

4.2 Producción de granos/frutas/hortalizas etc. en general desde Marzo 2006 hasta Febrero 2007

4.2.1 Cuál cultivos cultivó en su finca el año pasado? *Por favor, marca cuáles son*	4.2.2 Area cultivado en el año pasado? mz	4.2.3 Para que es utilizado? **(1) auto consumo** **(2) auto consumo y venta**	4.2.4 Cantidad cosechada en los últimos 12 meses? Quintal/mz	4.2.5 Cantidad vendida en los últimos 12 meses? Quintal/mz	4.2.6 Cuál fue el precio por quintal? C$/qq
a. Frijoles					
b. Maiz					
c. Frutas					
d. Guineo/platano					
e. Vegetales/Hortalizas					
f. Otro (especifica)					

PART 5: PRODUCCIÓN DE CAFÉ
5.1 Apectos generales del cafetal

5.1.1	Desde cuando trabaja en café?	(0) Desde joven (1) Hace ________ años (2) Desde ________ años
5.1.2	Cuántas manzanas de café tiene actualmente?	________ mz
5.1.3	Cuál es la distancia entre surcos (varas)?	
5.1.4	Cuál es la distancia entre plantas de café (varas)	

5.1.5	Cuántas manzanas de café tenía hace 5 años (2002)?	________ manzanas
5.1.6	Cuántas manzanas de café tenía hace 10 años (1997)?	________ manzanas
5.1.7	Si cultivó nuevo terreno con café en los últimos 10 años, qué tenía el terreno antes? (1) bosque natural (2) potrero (3) campo de siembra (4) café	
5.1.8	Qué edad tiene la parte más viejo de su plantación de café?	________ años
5.1.9	Cuántas mz tiene de café en desarrollo (de 1-3 años)	________ manzanas o ________ # de plantas

Administración / registro

5.1.10	Tiene registro de sus actividades y costos en la finca?	(0) No (1) Sí
5.1.11	Cuántas horas en la semana necesita para llenar el registro?	________ horas/semana

5.1.12 Costos para trabajadores

5.1.12.1	Contrata algún trabajador temporal de vez en cuando?	(0) No (1) Si
5.1.12.2	Si contrata un trabajador temporal, cuánto le paga al día?	________ C$/día
5.1.12.3	Cuántos veces le da comida al día?	________ #veces
5.1.12.4	Cuánto le cuesta una comida?	________ C$/comida
5.1.12.5	Tiene trabajadores permanente?	(0) No → (1) Si
5.1.12.6	Cuántas trabajadores permanente tiene?	________ # número
5.1.12.7	Cuánto le paga cada quincenal?	________ C$
5.1.12.8	Paga el séptimo?	(0) No (1) Si
5.1.12.9	Si paga el séptimo, cuánto le paga?	________ C$
5.1.12.10	Da comidas a los trabajdores permanentes?	(0) No (1) Sí
5.1.12.11	Si sí, cuántos comidas al día?	________ # número
5.1.12.12	Cuánto le cuesta una comida?	________ C$/comida
5.1.12.13	Dispone alojamiento para su trabajadores permanentes?	(0) No (1) Sí

Resiembra
5.1.13   Tenía semillero y almácigo/vivero en el año pasado?   (0) No   (1) Sí

5.1.13.1.1	Cuántas días hombres necesitaba en el año pasado para **establecer** el semillero y vivero *(incluye picado del terreno, llenar bolsas, siembra)*?	________ # trabajadores ________ # días
5.1.13.1.2	Cuántos de estos fueron contratados?	________ # trabajadores ________ # días
5.1.13.2.1	Cuántas tiempo necesitaba en los últimos 12 meses para **regar** el semillero y vivero?	________ horas al día ________ # de días
5.1.13.2.2	Cuántos de estos fueron contratados?	________ # trabajadores ________ # días

13

5.1.13.3	Cuántas bolsas compró para el vivero en los últimos 12 meses?	_________#cantidad
5.1.13.4	Cuánto le costó 1000 bolsas?	_________C$/1000 bolsas
5.1.13.5	Resembró en los últimos 12 meses un parte de su cafetal?	(0) No (1) Sí
5.1.13.6	Cuántos días/hombre utilizó para la resiembra?	_________# trabajadores _________# días
5.1.13.7	Cuántos de estos fueron contratados?	_________# trabajadores _________# días

5.1.14 Qué variedad de café cultiva?

	Marca cuál		*Marca cuál*	
a. Caturra		e. Maragojipe	
b. Bourbon		f. Typica	
c. Catuaí		g. Otra (especifica)	
d. Catimor			

5.1.15 Compró plantas jóvenes durante los dos años pasados?   (0) No   (1) Sí

	5.1.15.1 Cantidad comprado en 2006	5.1.15.2 Precio en 2006 C$/planta	5.1.15.3 Cantidad comprado 2005	5.1.15.4 Precio en 2005 C$/planta
a. Variedad _________				

5.1.16 Compró semillas durante los últimos dos años?   (0) No   (1) Sí

	5.1.16.1 Cantidad comprado 2006	5.1.16.2 Precio en 2006 C$/ kg semillas	5.1.16.3 Cantidad comprado 2005	5.1.16.4 Precio en 2005 C$/ kg semillas
a. Variedad _________				

5.2 Aplicaciones en el cafetal desde Marzo 2006 hasta Febrero 2007

Control de enfermedades (hojo de gallo, roya, mancha de hierro, pellejillo)

5.2.1	Ha realizado control de enfermedades en su finca en el año pasado?	(0) No (1) Si
5.2.1.1	Hizo la fumigación con químicos o con productos orgánicos?	(0) Químico (1) Orgánico

Si controló ENFERMEDADES con QUÍMICOS, por favor llena la siguente tabla

5.2.1.2	Qué producto ha utilizado? *Anota nombre por favor*	_______________
5.2.1.3	Cuántas aplicaciones en los últimos 12 meses?	_________# aplicaciones
5.2.1.4	Cuánto utilizo por bombada de 20 litros? (kg, l)	_________litros o kg/bombada
5.2.1.5	Cuantas bombadas por mz?	_________bombada/mz
5.2.1.6	Cuántas manzanas fumigó cada vez?	_________# mz
5.2.1.7	Costo del producto aplicado por unidad?	_________C$/por ____ *indica unidad*
5.2.1.8	Cuántas días hombre usó por aplicación?	_________# trabajadores _________# días
5.2.1.9	Cuántos de estos fueron contratados?	_________# trabajadores _________# días

14

Si controló ENFERMEDADES con PRODUCTOS ORGANICOS, llena la siguente tabla

5.2.1.10	Qué producto ha utilizado? (1) Sulfocalcio (2) Caldo bordelez (3) Caldo mineral vicosa (4) Caldo ceniza (5) Otro: *Anota nombre por favor*	_________________________
5.2.1.11	Cuántas aplicaciones en los últimos 12 meses?	____________ # aplicaciones
5.2.1.12	En cuántas manzanas aplicó el producto cada vez?	____________ manzanas
5.2.1.13	Cuánto litros utilizo por bombada de 20 litros?	____________ litros/bombada
5.2.1.14	Cuantas bombadas por mz?	____________ bombada/mz
5.2.1.15	Cuántas días hombre usó por aplicación?	____________ # trabajadores ____________ # días
5.2.1.16	Cuántos de estos fueron contratados?	____________ # trabajadores ____________ # días
5.2.1.17	Hizo el producto en la finca?	(0) No (1) Si
5.2.1.18	Si hizo el producto en la finca, cuánto tiempo demora elaborarlo?	____________ # trabajadores ____________ # días
5.2.1.19	Cuántos de estos fueron contratados?	____________ # trabajadores ____________ # días
5.2.1.20	Si compró el productó, cuánto le ha costado el litro?	____________ C$/por litro

Para la control de enfermedades con productos orgánicos:

	5.2.1.18.X. Qué material ha comprado?	5.2.1.19.X Qué cantidad necesitó por 1 mz?	5.2.1.20.X Costo por kg (C$)
(a) Azufre			
(b) Cal			
(c) Sulfato de cobre			
(d) Sulfato de zinc			
(e) Sulfato de magnesio			
(f) Otros (especifica)______			

Control de plagas (broca, minador, nemátodos, cochinilla etc.)

5.2.2	Ha realizado control de plagas en su finca en el año pasado?	(0) No (1) Si
5.2.2.1	Hizo la fumigación con químicos o con productos orgánicos?	(0) Químico (1) Orgánico

Si controló PLAGAS con QUIMICOS, por favor llena el siguiente tabla

5.2.2.2	Qué producto ha utilizado? *Anota nombre por favor*	_________________________
5.2.2.3	Cuántas aplicaciones en los últimos 12 meses?	____________ # aplicaciones
5.2.2.4	En cuántas manzanas aplicó el producto cada vez?	____________ manzanas
5.2.2.5	Cuánto utilizo por bombada de 20 litros? *(indica unidad)*	____________ litros/bombada
5.2.2.6	A cuanto lo compró la unidad? *(indica unidad)*	____________ C$/por ____
5.2.2.7	Cuantas bombadas por mz?	____________ bombada/mz
5.2.2.8	Cuántas días hombre usó por aplicación?	____________ # trabajadores ____________ # días
5.2.2.9	Cuántos de estos fueron contratados?	____________ # trabajadores ____________ # días

15

Si controló PLAGAS con productos ORGANICOS, por favor llena el siguente tabla

5.2.2.10	Qué producto ha utilizado? ***Anota nombre por favor** ej. Beauveria (HONGOS), sulforcalcio*	
5.2.2.11	Cuántas aplicaciones en los últimos 12 meses?	________ # aplicaciones
5.2.2.12	En cuántas manzanas aplicó el producto cada vez?	________manzanas
5.2.2.13	Cuánto litros utilizo por bombada de 20 litros?	_______litros/bombada
5.2.2.14	Cuanto bombadas por mz?	_______bombada/mz
5.2.2.15	Cuántas días hombre usó por aplicación?	_______ # trabajadores _______ # días
5.2.2.16	Cuántos de estos fueron contratados?	_______ # trabajadores _______ # días
5.2.2.17	Hace el producto organico en la finca?	(0) No (1) Si
5.2.2.18	Si compró el producto, cuánto le ha costado el litro o el kg?	_______C$/por unidad
5.2.6.3	Cuántos litros/ kg (de hongo / otros productos) compró en total para aplicarles el año pasado?	_____kg
5.2.2.19	Si hace el producto en la finca, cuánto tiempo demora elaborarlo?	_______ # trabajadores _______ # días
5.2.2.20	Cuántos de estos fueron contratados?	_______ # trabajadores _______ # días
5.2.2.21	Si hace producto en finca, cuál es el costo de los ingredientes por una manzana?	_______C$/por mz

5.2.3 Ha usado herbicidas en su cafetal en los últimos 12 meses? (0) No ➔ 5.2.4   (1) Sí

5.2.3.1	Qué tipo de herbicida ha utilizado? ***Anota nombre por favor***	______________
5.2.3.2	Cuántas aplicaciones en los últimos 12 meses?	_______ # aplicaciones
5.2.3.3	En cuántas manzanas aplicó el producto cada vez?	_______manzanas
5.2.3.4	Cuánto utilizo por bombada de 20 litros? (gramo, l ***indica unidad***)	________litros/bombada
5.2.3.5	Cuantas bombadas por mz?	_______bombada/mz
5.2.3.6	Costo de herbicida por unidad? (***indica unidad***)	C$/por ____
5.2.3.7	Cuántas días hombre usó por aplicación?	_______ # trabajadores _______ # días
5.2.3.8	Cuántos de estos fueron contratados?	_______ # trabajadores _______ # días

5.2.4 Ha usado ABONO SINTÈTICO en su cafetal en los últimos 12 meses? (0) No ➔ 5.2.5 (1) Sí

		Abono 1	Abono 2
5.2.4.1	Qué tipo de fertilizante ha usado (1) Abono foliar (2) Urea (3) Otro ***(anota por favor)***	☐	☐
5.2.4.2	Cuántas abonadas al año?	_______ # veces	_______ # veces
5.2.4.3	Cuántas manzanas abonó cada vez?	_____ # mz	_____ # mz
5.2.4.4	Cuánto aplicó por manzana? ***Indica si es quintal, litro o kg***	_______ por mz	_______ por mz
5.2.4.5	A cuánto le costó? ***Indica si es por quintal, litro o kg***	_______C$ por_____	_______C$ por_____
5.2.4.6	Cuántos días hombre usó por abonada?	_____ # trabajadores _______ # días	___ # trabajadores _______ # días
5.2.4.7	Cuántos de estos fueron contratados?	_______ # trabajadores _______ # días	___ # trabajadores _______ # días

16

5.2.5 Ha usado fertilizante orgánico en su cafetal desde Marzo 2006 hasta Febrero 2007?
(0) No ➜ 5.2.6   (1) Sí

5.2.5.1	Qué aplico? (1) Bocashi (2) Lombrihumus con pulpa de café (3) Composta (4) Abono foliar (4) Otros (especifica)___________________________ -	

Si aplicó **abono foliar orgánico:**

5.2.5.2	Qué tipo utilizó?	__________
5.2.5.3	Cuántas abonadas al año?	__________ # veces
5.2.5.4	Cuántas manzanas abonó cada vez?	__________ # mz
5.2.5.5	Cuánto aplicó por manzana?	__________ por mz
5.2.5.6	Cuántos días hombre usó por abonada?	__________ # trabajadores __________ # días
5.2.5.7	Cuántos de estos fueron contratados?	__________ # trabajadores __________ # días
5.2.5.8	Cuánto tiempo duró para elaborar el abono?	__________ # trabajadores __________ # días
5.2.5.9	Cuántos de estos fueron contratados?	__________ # trabajadores __________ # días

Si aplicó **otro fertilizante orgánico:**

5.2.5.10	Cuántas aplicaciones en los últimos 12 meses?	__________ # veces
5.2.5.11	Cuánto le hechó por planta cada vez?	__________ libra/planta
5.2.5.12	Cuantas manzanas fertilizó en total?	__________ mz
5.2.5.13	Cuántos qq de fertilizante compró en los últimos 12 meses?	__________ qq
5.2.5.14	Cuál fue el precio por qq?	__________ C$/qq
5.2.5.15	Que cantidad de fertilizante hizo en la finca?	__________ qq
5.2.5.16	Cuántas días hombre usó por aplicación?	__________ # trabajadores __________ # días
5.2.5.17	Cuántos de estos fueron contratados?	__________ # trabajadores __________ # días
5.2.5.18	Cuántas días hombre usó por preparación del fertilizante?	__________ # trabajadores __________ # días
5.2.5.19	Cuántos de estos fueron contratados?	__________ # trabajadores __________ # días

5.2.5.20 Si produjó fertilizante/abono foliar en la finca, qué ingredientes compró para hacerlo?

Qué material ha comprado?	5.2.5.21.X Qué cantidad?	5.2.5.22.X Viene en qué unidad?	5.2.5.23.X Precio de unidad (C$)
(a) Levadura			
(b) Cal			
(c) Estiércol/Gallinaza			
(d) Melaza			
(e) Carbón			
(f) Suero/leche			
(g) Sulfato de cobre			
(h) Sulfato de zinc			
(i) Sulfato de magnesio			
(j) Otros (especifica)______			

17

5.2.6 Utilizó trampas (con hormonas, alcohol, café) contra "la broca"?   (0) No   (1) Sí

5.2.6.1	Cuántas trampas utilizó en los últimos 12 meses?	_____ # trampas
5.2.6.2	Cuántas trampas compró nuevos?	_____ # trampas
5.2.6.3	Cuánto costó una trampa nueva?	_____ C$/ trampa
5.2.6.4	Cuánto costaron las hormonas/alcohol etc. por trampa?	_____ C$/llenar trampa
5.2.6.5	Cuántas veces renovó el liquido (las hormonas/alcohol)	_____ # veces
5.2.6.6	Cuántas días hombre duro la aplicar?	_____ # trabajadores _____ # días
5.2.6.7	Cuántos de estos fueron contratados?	_____ # trabajadores _____ # días

5.2.7 Dónde compra insumos necesario para la producción?

	5.2.7.1 **Por favor, indica de quíen**	5.2.7.2 Cuál es su modo de transporte? (1) Propio vehículo (2) Bestia, caminando (3) Vehículo prestado (4) Transporte público	5.2.7.3 A que distanci a son? km	5.2.7.4 Cuánto tiempo necesita para ida y vuelta? Horas	5.2.7.5 Si tiene que pagar por transporte, cuánto le cuesta (ida y vuelta)? C$/total	5.2.7.6 Cuántas veces al año compra insumos?
(1) Cooperativa						
(2) Bodega						
(3) Exportadora						
(4) Otro (especifica)						

5.3   Manejo del cafetal

5.3.1 Regó su cafetal desde Marzo 2006 hasta Febrero 2007? (0) No → 5.3.2   (1) Sí

5.3.1.1	Cuál es su fuente del agua? (1) Ojo (2) Río (3) Pozo (4) Quebrada (5) Lago	☐
5.3.1.2	Cuantós días hombre usó para regar el cafetal?	_____ # trabajadores _____ # días
5.3.1.3	Cuántos de estos fueron contratados?	_____ # trabajadores _____ # días
5.3.1.4	Como riega? (1) Motor de riego (2) Por gravedad	☐
5.3.1.5	Cuánto fueron los costos por el petroleo en los últimos 12 meses?	_____ C$

5.3.2 Hizo chapias en su cafetal desde Marzo 2006 - Febrero 2007?   (0) No ➜ 5.3.3   (1) Sí

5.3.2.1	Como hizo la chapia/deshierbe? (1) Manual con machete (2) Con herbicidas	☐
5.3.2.2	Cuántas chapias realizó en su cafetal?	_______ #
5.3.2.3	Cuántas días hombre usó para cada chapia?	_______ # trabajadores _______ # días
5.3.2.4	Cuántas de estos son contratado?	_______ # trabajadores _______ # días
5.3.2.5	Cómo pagó a los trabajadores contratados? (1) por día (2) por mz?	☐
5.3.2.6	Cuánto le pagó?	_______ C$

5.3.3 Realizó recepos en su cafetal desde Marzo 2006 - Febrero 2007? (0) No ➜ 5.3.4   (1) Sí

5.3.3.1	Como realizó el recepo? (1) Sistemático, alternando en bloques/lotes (2) Selectivo por planta	
5.3.3.2	Hace el recepo regularmente?	(0) No (1) Sí, cada_____años
5.3.3.4	Cuántas manzanas recepó en el último ciclo de producción?	_______ mz
5.3.3.5	Cuántas días hombre usó?	_______ # trabajadores _______ # días
5.3.3.6	Cuántos de estos fueron contratados?	_______ # trabajadores _______ # días

5.3.4 Hizo una poda sanitaria desde Marzo 2006 hasta Febrero 2007? (0) No ➜ 5.3.5   (1) Sí

5.3.4.1	Realiza la poda sanitaria regularmente?	(0) No (1) Sí, cada_____años
5.3.4.2	Cúantas días hombre usó para la poda?	_______ # trabajadores _______ # días
5.3.4.3	Cuántas de estos fueron contratados?	_______ # trabajadores _______ # días

5.3.5 Deshijó su café desde Marzo 2006 hasta Febrero 2007? (0) No ➜ 5.3.6   (1) Sí

5.3.5.1	Realiza la deshija regularmente?	(0) No (1) Sí, cada_____años
5.3.5.2	Cúantas días hombre usó para la deshija en el último ciclo de producción?	_______ # trabajadores _______ # días
5.3.5.3	Cuántas de estos fueron contratados?	_______ # trabajadores _______ # días

19

5.3.6 Manejo de sombra

5.3.6.1	Que tipo de sombra tiene? (1) Bosque rustico/natural (arboles viejos, 5 pisos) (2) Sombra tecnificada (3 pisos, ej. Inga)	
5.3.6.2	Cúantas arboles de sombra (incluyendo citrus y guineo) tiene en su cafetal?	__________ cantidad total
5.3.6.3	Hizó regulación de sombra en su cafetal **desde Marzo 2006 hasta Febrero 2007**?	(0) No (1) Sí
5.3.6.4	Cúantas días hombre usó para el manejo de sombra?	_________ # trabajadores _________ # días
5.3.6.5	Cuántas ce estos fueron contratados?	_________ # trabajadores _________ # días
5.3.6.6	Está reforestando la sombra de su cafetal?	(0) No (1) Sí

5.3.7 Tiene medidas de conservación del suelo? (0) No ➔ 5.3.8   (1) Sí

	5.3.7.1.X Cuáles medidas de conservación del suelo tiene en su cafetal? ***Marca por favor***	5.3.7.2.X Cúantas días hombre usó para mantener estas medidas?	5.3.7.3.X Cuántas de estos fueron contratados?	5.3.7.4.X Estado actual (1) bueno (2) regular (3) malo
a) Siembra en curvas a nivel		No llenar	No llenar	No llenar
b) Barreras vivas		____# trabajadores ____# días	____# trabajadores ____# días	
c) Barreras muertas		____# trabajadores ____# días	____# trabajadores ____# días	
d) Terrazas		____# trabajadores ____# días	____# trabajadores ____# días	
e) Cubertura del suelo		No llenar	No llenar	No llenar
f) Cortinas de rompeviento		____# trabajadores ____# días	____# trabajadores ____# días	
g) Otro (especifica)		____# trabajadores ____# días	____# trabajadores ____# días	

20

5.4 BENEFICIO HÚMEDO
5.4.1 El corte

5.4.1.1	**Practicó el graniteo / despinta?**	(0) No (1) Sí
5.4.1.2	Si sí, cuántas días hombre necesitaba para el graniteo?	__________ # trabajadores __________ # días
5.4.1.3	Cuántas de estos fueron contratados?	__________ # trabajadores __________ # días

5.4.1.4	Cuántas días duró el corte en total?	__________ # días
5.4.1.4.1	Cuándo **empezó el corte y cuándo terminó** ? | Desde__________	hasta__________
5.4.1.5	Cuántas latas cortó en total durante la cosecha?	__________ # latas
5.4.1.6	Hubo integración de su familia en el corte/cosecha de café?	(0) No (1) Sí
5.4.1.7	Cuántas latas cosechó/cortó su familia (sin estar pagado)?	__________ # latas
5.4.1.8	Cuánto pagó por una lata de café en el **corte principal**?	__________ C$/lata

5.4.1.9	Practicó repela?	(0) No (1) Sí
5.4.1.10	Si sí, cuántas días hombre necesita para la repela?	__________ # trabajadores __________ # días
5.4.1.11	Cuántas de estos fueron contratados?	__________ # trabajadores __________ # días
5.4.1.12	Practicó la pepena?	(0) No (1) Sí
5.4.1.13	Cuántas días hombre necesita para la pepena?	__________ # trabajadores __________ # días

5.4.1.14	Cuántas latas cortó en el graniteo?	__________ # latas
5.4.1.15	Cuántas latas cortó en la repela?	__________ # latas
5.4.1.16	Cuánto pagó por una lata de café en el grantiteo?	__________ C$/lata
5.4.1.17	Cuánto pagó por una lata de café en la repela?	__________ C$/lata

5.4.2 Hizo el despulpado en su finca (o un colectivo privado)? (0) No ➜ 5.4.3 (1) Sí

5.4.2.1	Cuánto tardó en despulpar 20 latas?	__________ horas
5.4.2.2	Quién despulpó? (0) Entrevistado o miembros de su familia (no pagado) (1) Contrató a alguien	☐
5.4.2.3	Si contrató, cuánto le pagó por 20 latas?	__________ C$/lata
5.4.2.4	Cuántos días hombre dura para lavar y secar el café que sale de las 20 latas?	__________ # trabajadores __________ # días
5.4.2.5	Quién lo hizó? (0) Entrevistado o miembros de su familia (no pagado) (1) Contrató a alguien	☐
5.4.2.6	Cuánto le ha pagado al contratado por día?	__________ C$/día
5.4.2.7	Mientras se seca el café, hace el escojido?	(0) No (1) Sí
5.4.2.8	Dónde lavó el café? (1) En río, quebrada (2) Cerca del río, quebrada, crique *(anota a que distancia)* (3) Lejos del quebrada, crique en el beneficio	☐

21

5.4.2.9	Qué hizo con los aguas mieles? (1) Depositado en río / quebrada (2) Depositado en la tierra (3) Laguna para sedimentación natural (4) Fosas especiales de infiltración (5) Pilas deconcretos	
5.4.2.10	Estába aprovechando el mucílago/agua miel? (0) Nada (1) Herbicida (2) Mezcla con la abonera (3) Abono foliar	
5.4.2.11	Qué hizo con la pulpa de café? (0) Botarla (1) Procesarla en abono (2) Alimentación de ganado	
5.4.2.12	Tenía gastos en el beneficio humedo por electricidad o gasolina?	(0) No (1) Si
5.4.2.13	Si tenía gastos, cuánto fueron los gastos en total?	___________ C$

5.4.3 Si no hace el despulpado en su finca

5.4.3.1	Si no lo hace en propia finca, a dónde lleva su café en uva para el despulpe? (1) Despulpadora de la cooperativa (2) Vecino (3) Otro (especifica)___________________	_________________
5.4.3.2	Cuánto tenía que pagar por el servicio por carga? ***Indica unidad por favor***	___________C$/carga
5.4.3.3	Cuánto fueron los costos de transporte por carga uva para llevarlo al beneficio húmedo?	___________C$/carga
5.4.3.4	Tenía otros costos adicionales? Cuántos?	___________ C$

5.4.4 Reparaciones y herramientas para manejo del cafetal y beneficio

5.4.4.1	Cuántos fueron los costos de reparación de la despulpadora mientras el corte de 2006/7	___________ C$
5.4.4.2	Cuántos fueron los costos de reparación del beneficio antes/mientras el corte de 2006/7	___________ C$
5.4.4.3	Cuenta con todos los herramientas necesario para el manejo del cafetal y el beneficio húmedo?	(0) No (1) Si
5.4.4.4	Si no, cuántos fueron los costos totales para las herramientas que pidió prestado para el manejo del cafetal y de beneficio húmedo en los últimos 12 meses?	___________ C$
5.4.4.5	Cuántos sacos compró para el corte 2006/2007?	_______ # número
5.4.4.6	Cuánto costó un saco?	_______ C$/saco
5.4.4.7	Cuántas canastas compró para el corte 2006/2007?	_______ # número
5.4.4.8	Cuánto costó una docena?	_______ C$/docena

Por favor, especifique en que forma entregó el café y para que le pagaron (si pagaron por carga oreado o por quintal oro...?)

5.4.5 A dónde vende su café?	5.4.5.1.X Cuántas **cargas o qq** vendió a...? *Indica unidad por favor*	5.4.5.2.X En que forma entregó el café? (1) Café en uva (2) pergamino oreado" (42% humedad) (3) húmedo (44-48%) (4) mojado (56%) (5) oro	5.4.5.3.X Cuántas veces entrego el café durante el último corte al ....?	5.4.5.4.X Se hace una liquidación final o por cada entrega? (1) Liquidación final (2) Por cada entrega	5.4.5.5.X *Si es una liquidación final:* Cúal fue el precio que recibió? *Anota si es oreado o oro y por carga o por qq!*	Si los precios son diferentes con cada entrega, anote cuántas cargas o qq y por qué precio vendió cada una!			
						5.4.5.6.X **Cantidad de la primera entrega**	5.4.5.7.X Cuál fue el precio por carga/qq en la primera entrega? C\$/carga o qq	5.4.5.8.X **Cantidad de la segunda entrega**	5.4.5.9.X Cuál fue el precio por carga/qq en la segunda entrega? C\$/carga o qq
(0) La cooperativa dónde esta socio									
(1) Casa comercializadora (CISA/Exportador Atlantic...)									
(2) Intermediario/acopio fuera de la finca									
(3) Intermediario que viene a la finca									
(4) Otra cooperativa con beneficio seco y comercialización									
(5) Otro *(especifica)________*									

NOTA: ES MUY IMPORTANTE ANOTAR CUÁNTO CAFÉ VENDIÓ EN TOTAL Y A QUE PRECIO -

	5.4.5.10.X Cómo transporta su producto al punto de acopio **(Código 11)**	5.4.5.11.X El medio del transporte es (1) Propio (2) Prestado (3) Compartido (4) Alquilado (5) Cooperativa (6) Público	5.4.5.12.X Cuánto tenía que pagar por el transporte en total?	5.4.5.13.X A que distancia es el acopio de la finca? In km o horas/minutos	5.4.5.14.X Cuánto tiempo necesita para llevar el café por alla? horas/minutos
(0) La cooperativa dónde esta socio					
(1) Casa comercializadora (CISA, Exportador Atlantic)					
(2) Intermediario/acopio fuera de la finca					
(3) Intermediario viene a la finca	No llenar	No llenar	No llenar	No llenar	No llenar
(4) Otra cooperativa con beneficio seco y comercialización					
(5) Otro ___________					

Código 11
(1) Camioneta, Camión
(2) Carreta
(3) Bestia
(4) Viene a recoger los productos
(5) A pie
(6) Transporte público

24

5.4.5 Cantidad de cosecha

5.4.5.1	Cuántos qq de café oro/uva/pergamino ha cosechado en la cosecha del año antepasado **(2005/6)**? *Por favor, indica que tipo de café entregó*	_____________ qq
5.4.5.2	Cuál fue el precio que recibió por su cafe qq café oro/uva/pergamino el año antepasado (cosecha 2005/6)? *Por favor, indica que tipo de café entregó*	_____________ C$/qq

5.4.6 Pago de café

5.4.6.1	De dónde recibe información sobre el precio del café? (1) Cooperativa (2) Radio (3) Periódico (4) Televisión (5) Amigos/Familia etc. (6) Exportadora (7) Acopio (8) Otro_____________	☐
5.4.6.2	Como financió el ciclo de producción desde Marzo 2006 hasta Febrero 2007? (0) Vende el café de futuro a la cooperativa (1) Crédito de la cooperativa (2) Vende el café de futuro al exportadora (3) Crédito del banco (4) Crédito de vecino/intermediario (5) Prestamo de la familia (6) Propio ahorros (7) Otro	☐
5.4.6.3	En que fecha entregó su último cosecha a la cooperativa?	_____________ fecha
5.4.6.4	Cúando recibió su liquidación final?	_____________ fecha
5.4.6.5	Si dice que todavía no lo recibió, cuándo espera su liquidacion final?	_____________ mes
5.4.6.6	Cómo financia el tiempo entre el último corte y la liquidación final? (0) Uso de un adelanto sin interés (1) Uso de ahorros (2) Tomar nuevo crédito	☐
	Si toma un nuevo crédito/prestamo para cubrir los gastos hasta que recibe la liquidación final:	
5.4.6.7	Cuánto C$ fue el nuevo crédito que solicitó?	_____________ C$
5.4.6.8	Por cuántos meses?	_____________ meses
5.4.6.9	Cuál es el interés? (*o cuánto tiene que repagar en total)*?	_____________ % interés
5.4.6.10	Por favor especifique dónde solicitó el crédito : (1) Banco (2) Exportador (3) Acopio (4) Cooperativa (5) Prestamista (6) Familiar/ Amigo (7) ONG/proyecto de desarrollo (8) Otro (especifique)	

25

5.5 Certificación de café   SÓLO PARA PRODUCTORES CON CERTIFICACIÓN:

5.5.1 Tipo de Certificatión

	5.5.1.X.1 Su café tiene la certificación de.....? (0) No (1) Sí (2) En transición	5.5.1.X.2 En qué año recibió la certificación?
a. Café orgánico - OCIA		
b. Café orgánico BIOLATINA		
c. Café orgánico- IMO/NATURLAND		
d. Comercio justo		
e. Otro________		

5.5.2 Costos de certificación

5.5.2.1	Quién hace el trabajo administrativo relacionado con la certificación? (1) Cooperativa (2) Exportador	
5.5.2.2	Cuántas días dura la inspección interna al año en su finca?	_______ # días
5.5.2.3	Tiene que pagar cada año para la certificación?	(0) No (1) Sí
5.5.2.4	Si sí, como paga por la certificación orgánica? (1) por quintal café oro producido (2) por manzana (tamaño de finca) (3) el mismo para todos productores (independiente del tamaño de la finca)	
5.5.2.5	Si tiene certificación de **comercio justo y organico**, como paga por las certificaciónes? (1) por quintal café oro producido (2) por manzana (farm size) (3) el mismo para todos productores (independiente del tamaño de la finca)	
5.5.2.6	Cuánto tiene que pagar para la certificación cada año? *(indica lo que aplica)*	**Para sello organico:** _________ C$ total o _________C$/qq o mz **Para sello organico-comercio justo juntos** _________ C$ total or _________C$/qq o mz
5.5.2.7	***Solo si tienen también COMERCIO JUSTO:*** Conoce el precio minimo del café con certificación orgánico - comercio justo?	(0) No (1) Sí, ___________U$/qq oro
5.5.2.8	Recibe un premio por el café orgánico / orgánico/comercio justo?	(0) No (1) Sí, ___________C$/qq oro/pergamino
5.5.2.9	Porque produce orgánico? (1) Protección del medio ambiente (2) Mejor por la salud de la familia (3) Más capacitacion (4) Más actividades sociales que benefician a familia, comunidad (5) Ingreso mejor y más alto (6) Otro (especifa________	

26

5.5.5 Antes/después de la certificación

Actividad	5.5.5.X.1 Para los siguientes actividades, **los costos de producción** ahora son iguales con certificación que sin certificación? (1) Lo mismo que antes (2) Más que antes (3) Menos que antes	5.5.5.X.2 Puede indicar el porcentaje cuánto los costos de producción fueron más o menos? %
a. Deshierbe/Chapia		
b. Fertilización		
c. Control de plagas y enfermedades		

Actividad	5.5.5.X.3 Para los siguientes actividades, necesita ahora con certificación la misma **mano de obra** que sin certificación? (1) Lo mismo cantidad que antes (2) Más que antes (3) Menos que antes	5.5.5.X.4 Puede indicar el porcentaje cuánto el mano de obra fue más o menos? %
a. Deshierbe/Chapia		
b. Fertilización		
c. Control de plagas y enfermedades		

5.5.5.5	Había una baja de rendimiento por mz cuándo ha cambiado de producción tradicional/convenciónal al orgánico?	(0) No (1) Sí
5.5.5.6	Como fue su sistema de producción antes de la certificación? (1) Tradicional sin agroquímicos (2) Tradicional con algunos agroquímicos (3) Semi-intenso	
5.5.5.7	El rendimiento por mz ahora está ? (1) Menos que antes de la certificación (2) Igual como antes (3) Más que antes **Por favor, indica % más o menos o cuantas cargas sacó antes y ahora**	________% más o menos que antes

5.5.5.8 Imaginese que en el travesaño del medio representa su <u>nivel de vida</u> antes de tener la certificación, en cual travesaño de esta escalera ubican ahora el nivel de vida?

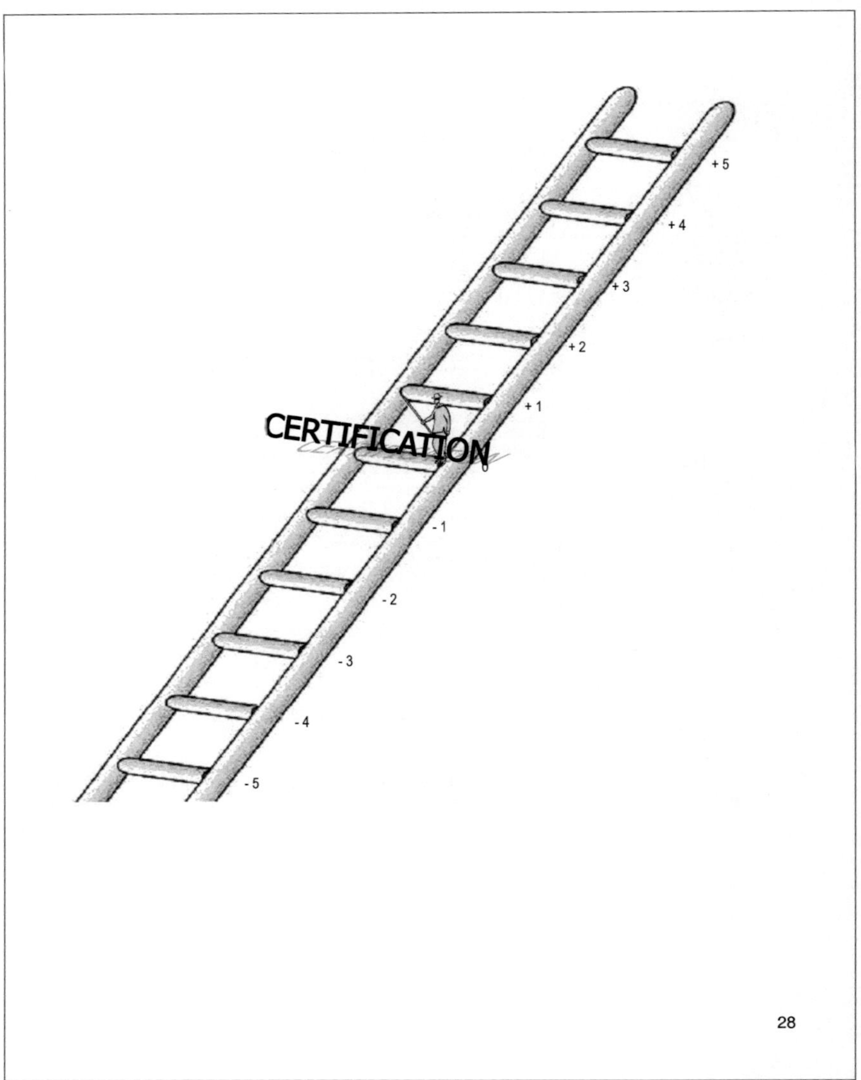
+ 5
+ 4
+ 3
+ 2
+ 1
CERTIFICATION
- 1
- 2
- 3
- 4
- 5
28

PART 6: SOCIO EN ORGANISACIONES Y CAPITAL SOCIAL

6.1 Socios en organisaciones, asociaciones y grupos aparte de la cooperativa por cuál vinimos
(Pregunta lo siguiente por todos los miembros del hogar MAYOR DE 15 AÑOS, llenalo por todo los organisaciones el hogar participa. Si a miembro participa en varias organisaciones, entrega su ID en la primera columna varias veces and llena una linea por cada organización la persona es socio).

NO LLENAR POR LA COOPERATIVA DE CUAL IDENTIFICAMOS AL ENTREVISTADO➔ está información sale en la próxima página

6.1.1 ID del miembro del hogar (usa ID de la página 3)	6.1.2 Typo de organisación (Código 12	6.1.3 Grado de participación (Código 13)	6.1.4 En los últimos 12 meses, hizo contribuciones a está organisación? (1) En efectivo, **cuánto C$?** (2) En bienes, servicios (e.g. labor, etc.)	6.1.5 Participa regulariamente en las asambleas/reuniones? (0) No (1) Sí (2) De vez en cuando (3) No hay asambleas

Código 12 Typo de organización (por favor, especifica nombre)

(1) Cooperativa de servicios multiples
(2) Cooperativa de ahorro y credito
(3) Uniones (UNAG, UNICAFE)
(4) Microfinanciera
(5) Grupo ambientalista
(6) Miembro de un partido politica
(7) Grupo religioso
(8) Consejo de padres de familia
(9) Grupo de jovenes
(10) Grupo de deporte
(11) Otra (especifica)

Código 13 Grado de participación

(1) Directivo
(2) Otro cargo
(3) Activo
(5) No activo

29

6.2 Relacionado a "su" cooperativa *(con cúal estamos cooperando)*

6.2.1	Desde cuándo es socio de está cooperativa?	_____año
6.2.2	Cuánto es el aporte anual a la cooperativa?	_____C$
6.2.3	Cargo en la cooperativa (1) Presidente/Vice-presidente (2) Muy activo (otro responsabilidad, en la junta...) (3) Delegado para la cooperativa del segundo grado (4) Solo socio	
6.2.4	Participa regularmente en las asambleas/reuniones extraordinaria? (0) No (1) Sí (2) De vez en cuando (3) No hay asambleas	
6.2.5	Cuántas veces en los últimos 12 meses participa en reuniones de la cooperativa?	_____# número
6.2.6	Cuánto tiempo dura aproximadamente una reúnión (incluso su transporte?)	_____horas/ reunión
6.2.7	Reciben capacitaciones para la producción del café? *(incluye capacitaciones de la cooperativa de base y de segundo grado)*	(0) No (1) Sí
6.2.8	Si sí, cuantas en los últimos 12 meses	_____# número
6.2.9	Cuanto tiempo duró cada uno (incluso transporte)	_____horas/ reunión
6.2.10	Tiene gastos para e transporte, la comida o otros cosas para las reuniones? Si sí, cuánto por reunión?	(0) No (1) Sí, _____C$/reunion
6.2.11	Porque es socio de la cooperativa de base? (0) Inseguridades con la tierra (1) Aceso a créditos (2) Aceso a información/capacitación (3) Comercialisación del café (4) Mejores precios para el café / precios más estables (5) No hay otra solución porque la cooperativa está endeudada (6) Unica manera para aceso a certificación (7) Otro, especifica__________________	
6.2.12	De cuáles actividades de su cooperativa o de la de segundo grado ha sido beneficiado usted o un miembro de su hogar? (1) Becas para estudiar (2) Material para estudiantes (lapiz, cuadernos) (3) Construción de caminos, edificios, escuelas (4) Apoyo a grupos (de jovenes, sports...) (5) Mejoramiento de la situación medicinal (6) Mejoramiento del medio ambiente (7) Otro (especifica_________________)	

6.3 Contactos de información

6.3.1	De quién recibe información sobre producción de café y capacitaciones? (0) Nadie (1) Técnicos de cooperativa/directivos (2) Técnicos de la cooperativa de segundo grado (3) Exportador (4) Beneficio (5) Magfor (6) Unicafe (7) Radio (8) Periodicos (9) Televisión (10) Otro________	

30

6.4.1. Cuantas personas conoce que **trabajen** con las siguientes organizaciones?

Tipo de organización	6.4.1.1.X. Cuantas personas conoce?	6.4.1.2.X. Cuantas personas de estos son familiares?	6.4.1.3.X Cuantas personas son amigos de algún miembro de la familia?
a. Cooperativas agrícolas o crediticias *(aparte de su cooperativa)*			
b. MAGFOR			
c. Uniones (UNAG, UNICAFE...)			
d. Beneficio seco/ Exportadores de café			
e. Organisaciones con proyectos de desarrollo			
f. Partido Pólitico			

6.4.2. En caso de un problema o emergencia como los listados abajo, sería fácil o no de pedir ayuda de alguna de las siguientes personas o de su cooperativa?
(1) Fácil
(2) Díficil

Tipo de problema/emergencia	Pariente cercano	Pariente lejano	Amigo/vecino	Cooperativa
a. Pedir prestado dinero para educación				
b. Pedir prestado dinero para gastos médicos				
c. Pedir prestado dinero para una celebración				
d. Pedir prestado dinero para alimentación				
e. Pedir ayuda para trabajar en la finca				

6.4.3 Por favor indique si usted esta de acuerdo o en desacuerdo con las siguientes afirmaciones:
(1) De acuerdo (sí)
(2) En desacuerdo (no)

Entrevistadora, lea estas frases lentamente al entrevistado:

a.	La mayoría de las personas en la comunidad son honestas y en ellas se puede confiar.	
b.	Las personas de esta comunidad estan solamente interesadas en su propio beneficio.	
c.	En caso de tener problemas, siempre hay alguna persona que le ayude.	
d.	Si un animal se le pierde (cerdo, caballo, vaca), alguien en la comunidad le ayudará a buscarlo o le lo regresará.	

31

PART 7. ESCALERAS

Entrevistadora: por favor mostrar al entrevistado la foto de la escalera de diez travesaños.
7.1. Esta es una foto de la escalera de diez travesaños. Imaginese que en el travesaño de abajo, primero, se encuentran las personas mas pobres, y en el mas alto, último se encuentran las personas mas ricas del municipio. En cual travesaño de esta escalera esta ubicada su familia?

7.2. En cual escalón de esta escalera, podria usted colocar a una familia (esposo, esposa, 4 hijos) cuyo ingreso mensual es igual a 2500 C$?

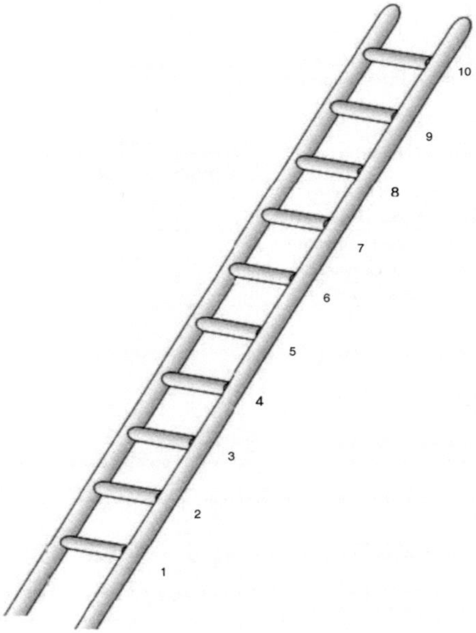

32

7.3 Esta foto muestra una escalera. Suponiendo que el escalón en el medio representa sus condiciones de vida en el año 2002, en cual escalón piensa usted que se encuentra en los actuales momentos?

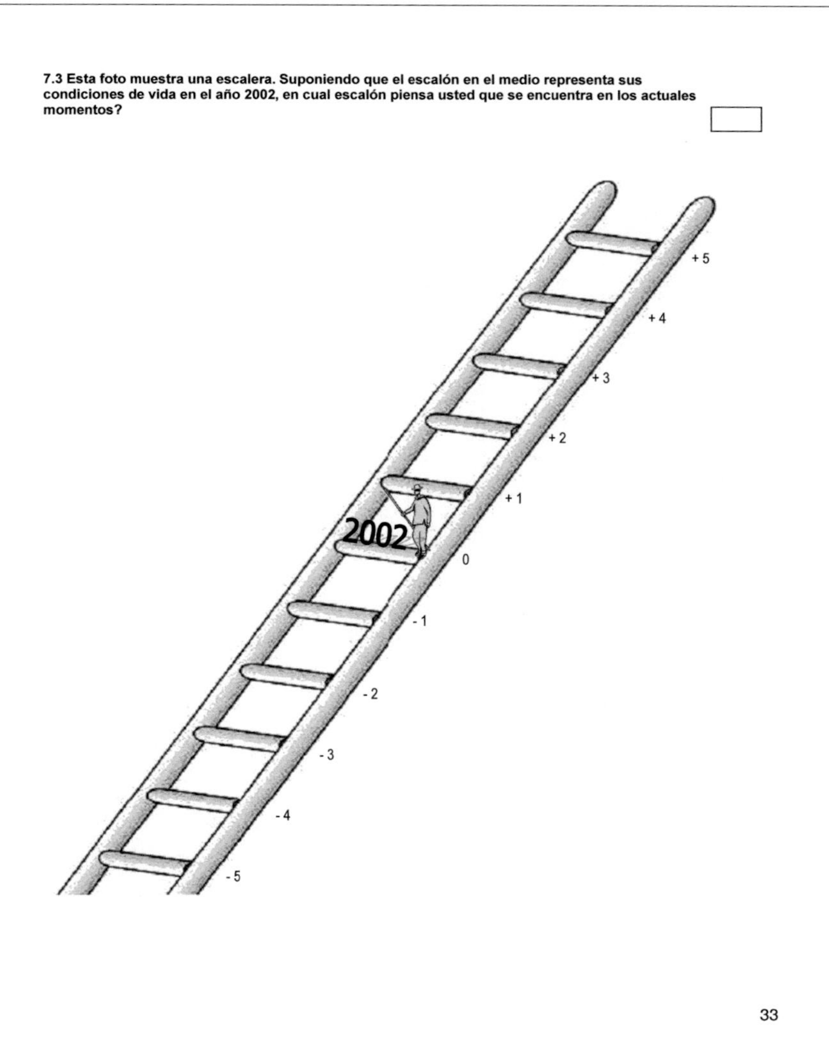

PART 8: ACCESO A SERVICIOS FINANCIEROS Y CRÉDITOS
Para ser llenado por los miembros de la familia mayores de 15 años   * ➔Mas de una (1) respuesta es posible!!!

No.	8.1 Ha solicitado alguna vez un crédito/adelant o/prestamo? (0) No (1) Si	8.2 **Si no ha solicitado crédito,por que no?** **código 14**	8.3 A partir del Marzo 2006, cuántos créditos/prestamos/venta a futuro ha recibido? *(anota el meses y año)*	8.4 A que institución /persona solicitó el crédito? **código 15** *(varias respuestas son posible, incluyendo créditos informales provenientes de familiares ...)*	8.6 Cuál fue el monto de cada crédito desde marzo 2006 hasta ahora? C$	8.7 Cuál garantía fue solicitada para los créditos? **código 16***
01			1 2 3 4	1 2 3 4	1 2 3 4	1 2 3 4
02			1 2 3		1 2 3	1 2 3
03					1 2	1 2
04						
05						

CÓDIGO 14

(1) Suficiente dinero propio
(2) Ingreso estimado del café es muy bajo para la inve
(3) No posee suficientes garantias
(4) Condiciones desfavorables
(5) Procedimiento muy complicado
(6) No sabe como solicitarlo
(7) No puede leer/escribir
(8) Temeroso de las deudas
(9) El banco esta muy alejado
(10) No sabe porque el crédito fue rechazado
(11) Otros, especifique:

CÓDIGO 15

Por favor especifique exactamente de quien:
(1) Banco
(2) Exportador
(3) Acopio
(4) Cooperativa **(anota cuál)**
(5) Prestamista
(6) Familiar/ Amigo
(7) ONG/proyecto de desarrollo
(8) Otro (especifique)

Código 16

(1) Ninguno
(2) Escritura de propiedad
(3) Carta de venta de animales
(4) Bienes de consumo
(5) Fiador
(6) Venta de futuro de café
(7) Otros (especifique):_______

8 ACCESO A SERVICIOS FINANCIEROS Y CRÉDITOS (continuación)

Nota: Para ser llenado por los miembros de la familia mayores de 15 años   * → **Mas de una (1) respuesta es posible!!!**

No	8.8 Tenían los créditos diferentes tasas de interés? (0) No (1) Si	8.9 Si los créditos tenían la misma tasa de interés, cual fue dicha tasa? % mensual o % anual *(por favour indique cual)*	8.10 Si tenían un interés diferente, cuál fue la tasa de **interés** para los créditos? *o para facilitar:* **Cuanto es el monto que tiene pagar en total?**	8.10.1 Si no está incluido en el interés, cuánto fue el porcentaje por el **mantenimiento del valor?**	8.10.2 Si no está incluido en el interés, cuánto fue el porcentaje por la **administración?**	8.13 En qué utilizo el crédito? Código 17*	8.14 Puede pagar el crédito a tiempo? (0) No (1) Si (2) Quizás (3) Ya cancelado	8.14.1 En que fecha pagó o tiene que pagar el crédito?	8.15 Si usted no puede o no está seguro de poder pagar el crédito al tiempo → por que? Código 18*
01			1 2 3 4	1 2 3 4	1 2 3 4	1 2 3 4	1 2 3 4	1 2 3 4	1 2 3 4
02			1 2 3	1 2 3	1 2 3	1 2 3	1 2 3	1 2 3	1 2 3
03			1 2	1 2	1 2	1 2	1 2	1 2	1 2
04									
05									

CÓDIGO 17

Alimentación... (1)
Gastos médicos .... (2)
Educación.. (3)
Matrimonio, funeral, otras ceremonias......... (4)
Pago de trabajadores de la finca .... (5)
Comercialización del café .... (6)
Insumos para el café (fertilizante, pesticidas)(7)
Insumos para otros cultivos (semillas, fertilizante, pesticidas) .... (8)
Construcción (casa, deposito)...................... (9)
Otros_________________________ .. (10)

CÓDIGO 18

(1) Falta de dinero	(9) Compra de bienes de consumo
Muy altos costos por	(10) Pago a prestamista
(2) - Gastos médicos	(11) Regalo dinero a familia/amigo .
(3) - Educación	(12) Intereses son muy altos
(4) - Matrimonio, funeral, otros	(13) Otros. Especifique
(5) - Insumos	_________________
(6) - Trabajo de construcción	
(7) No pudo vender la cosecha	
(8) Mala cosecha	

8 ACCESO A SERVICIOS FINANCIEROS Y CRÉDITOS (continuación)

Nota: Para ser llenado por los miembros de la familia mayores de 16 años   * ➔Mas de una (1) respuesta es posible!!!

	8.18 Tiene todavía deudas por créditos que haya adquirido antes del marzo 2006? (0) No (1) Si	8.19 Comparando el monto de sus deudas actuales con el valor total de sus bienes (incluyendo bienes, animales, mueble), su deuda actual es: (1) Menos de la mitad (2) La mitad (3) Mas de la mitad (4) No sabe	8.20 Tiene usted alguna cuenta bancaria? Por ejemplo de ahorro o corriente? (0) No (1) Si
01			
02			
03			
04			
05			

> ➢ Decir gracias a los productores
> ➢ Informarles que con una gran posibilidad va a venir algúien mientras es la epoca de cosecha quién va a tomar una muestra de café para que so puede relacionar estos datos con datos de calidad de café.

Cooperative Questionnaire

**Análisis de participación en las cadenas de
comercialización del café certificado y sus efectos
en el bienestar de los pequeños agricultores**

Encuesta al nivel de la cooperativa

Fecha:

Cooperativa:

Nombre y appelido del entrevistado/a:

Cargo en la cooperativa:

1

En general:

1	Año de fundación de la cooperativa:
2	Quien la fundó? El gobierno? (=> UCAS?)　　　ONG　　　Productores?
3	➜ Cuentá brevemente la historia por favor (etapas importantes):

Cuáles son los objetivos, la vision y la mision de la cooperativa?	
objectivos	
visión	
misión	

Cuál es el actividad principal como cooperativa?
Cuáles servicios ofrece a sus socios y non-socios?

Servicios ofrecidos	Socios	Non-socios	
Compra y venta de insumos			
Créditos			
Compra y venta de café			
Asistencia legal			
Tienda de cooperativa			**Cantidad/año**
Talleres sobre producción del café			
Talleres sobre producción agricola (diversificación)			
Talleres sobre manejo economico de la finca/negocios			
Visita regulares de tecnicos a los productores			
Otros (especifique)			

Créditos

Cuál es la taza de interés para los créditos?
Cuanto cobra por el mantenimento de valor?
Cuaél es la taza de administración?
Cuál es el periódo máximo que dan un credito?

Socios:

Número total de socios
Número de socios que tienen café
Desarrollo de los socios en los últimos 5 años? Bajando　　Constante　　Creciendo poco　　Creciendo mucho **Razones:**
Hay condiciones para asociarse? No　　　　Sí, cuáles (ej. educacion, area de finca, cultivo, edad, genero??

	Cuánto es el pago de membrecía?
	Qué pasa con el pago de membrecía si uno sale de la cooperativa? Recibe más dinero por la inversion de la cooperativa?
	Mz totales que tienen los socios de la cooperativa:
	Mz totales de café:
	Ø Mz/café por productor
	Ø del rendimiento de qq/mz de café:
	Cómo se desarrollo el rendimiento qq/mz en los últimos 5 años? Razones:

Estructura

	Cuál es la estructura de la cooperativa (organigrama)?
	Comité Ejecutivo de Dirección (Vorstand) Consejo de Administración Consejo/Junta de Vigilancia Gremios de crédito Gremio de educación Gremio del control interno
	Cuantos gremios tiene con cuántas personas? Reciben un pago?
	Cuántas mujeres tienen un cargo en la cooperativa?
	Un socio – un voto??

Gerencia:

	Cuántos empleados tiene en total?
	• Administradoras?
	• Certificación (control interno)?
	• Tecnicos? (tecnicos financed from whom, how did they get that project?)
	• Que desembolsa los creditos?
	• Promotores de crédito?
	• Tiene un escecialista de venta/exportación? O hace gerente?
	Cuántos son mujeres? Porqué?
	Tiene una alta fluctuación de los empleados?
	Está contento con el nivel de calificación de sus empleados?
	Tiene incentivos para premiar sus empleados buenos?
	Tiene problemas en conseguir empleados buen califacados?
	Cuál carrera hizo el/la gerente?
	Cuál es el nivel educativo del presidente de la cooperativa?

Recursos y Economía:

	Cual recursos tiene la cooperativa? Computadoras Generador GPS Motos Carros Camiones
	Quién financió la casa de la cooperativa? Cuándo?
	Tiene un beneficio seco?

3

Cuántos acopios tiene?   Distancia hacia el beneficio que utiliza?
Tiene un plan de negocios por los próximos 3-5?   ☐ Sí   ☐ No   ☐ Inseguro
Cuál describe major su estrategia de negocios?   ☐ Crecimiento de coop/venta   ☐ Estabilidad   ☐ Cambio estrategico
Hace la contabilidad a mano o con computadora?
Cuál es el mayor ingreso de su cooperativa?
La cooperativa hace ganancias cada año?
Tiene un fondo de reserva/ capitalización?
Cómo se financia la cooperativa? (y la cosecha...)?
La coop tiene aceso a créditos? De quién?   Corto-plaza (1 año)   Largo-plazo
Cuál es la taza de interés?
Tiene suficiente créditos o necesitará más?   A qué condiciones?
Le cuesta de pagar los créditos, especialmente los de largo plazo?   ☐ Sí   ☐ No   ☐ Inseguro
La situación financiera de la cooperativa se mejoro en los últimos 5 años?   ☐ Peor   ☐ Constante   ☐ Poco mejor   ☐ Mucho mejor
Qué hace con sus ganancias cuáles genera la cooperativa?   Lo distribuye? Cómo?
Cuáles inversiones hizo la cooperativa en los últimos 10 años? Quién financió?
Puede estimar los costos en un año para ?   a) todos talleres relacionado a producción de café: b) del servicio tecnico?:
Cómo financia los talleres para productores?

<u>Compra de café de los productores</u>

Sabe cuánto son los costos para producir un qq de café? Y de café orgánico?
Paga un adelanto sin interés para la cosecha de los productores?   Como se calcula cuanto da a un producto?
Cómo organiza la compra de café? (Tareas de la coop de base/ segundo grado)
Tiene un contrato con los productores?   Tambien arregla cuanto café tienen que llevar?   Los precios se definen cuando?
Paga alta calidad por los productores? Cómo?
Castiga por baja calidad? Cómo?

4

Cuál son los costos del precio final que hay deducir a los productores por el procesamiento, exportación, etc.?

- o Costos del procesamiento por qq?_______________________________
- o Costos del exportación por qq?_______________________________
- o Costos por administración?_______________________________
- o Costos por pago de deudas de la coop?_______________________________
- o Costos por fondo de capitalización?_______________________________
- o Costos por certificación?_______________________________
- o Otros costos, especifique?

Existe competencia entre su cooperativa y otras cooperativas, coyotes, exportadores?

□ No   Sí, especifica de quién....

Cuál es la intensidad de la competencia de compradors de café?

□ Muy intenso   □ Intenso   □ Algo intenso   □ No intenso

Está facil por un comprador de café entrar en el negocio de café?

□ Muy facil   □ Facíl   □ Algo facíl   □ Muy difícil

Cuál son las ventajas y desventajas de su competencia? ➔ Porqué venden los productores alla?

Cómo reaccionan sus miembros a la competencia? Salen de la cooperativa actualmente?

Tiene problemas con no cumplir con los contratos de los productores?
Puede estimar una taza de cuantos no cumplen?
Cuál son razones?
Que hace entonces?

Cuál es la intensidad de la competencia en el mercado de exportación?
Muy competitivo   Competitivo   Algo competitivo   No competitivo

---

Separa café según la calidad?
Y lo vende separado?

Si vende de alta calidad, cómo lo distribuye las ganancias a los productores-solo los que lo producen o igual a todos?

Venta de café

En los últimos 5 años, la cooperativa desarrollo una nueva linea de café ej. Café de altura/calidad; nueva certificación?

□ No   □ Si, cuál?

En los últimos 5 años, la cooperativa mejoro la calidad de su café vendido?

□ No   □ Si, cómo?

Tiene su propia marca que vende?
Dónde lo vende?
Qué cantidad?

Cuanto café vendio en total en 2006-7?

Podría decir el desarrollo en las ventas de café en los últimos 5 años?

5

A quien vendió el café en 2007? Desde cuándo vende a el? Qué tipo y cantidad? Que precio por qq?		
Comprador /desde cuando	Tipo café (cert) y cantidad	Precio/qq oro?
Le pagan un premio por calidad? Cuánto?		
Siempre pueden cumplir con los contratos de los importadoras??		
Por cuanto tiempo son los contratos? ☐ un ciclo ☐ Contratos de largo plazo, DE QUIÉN?		
Cuánto café vende con contratos de largo plazo?		
Sabe porque los importadores compran de su cooperativa?		

Red social:

	Con cuál de los siguientes trabajó en el pasado? Dónde son basado? ONGs Cooperativas Organizaciones (UNAG...)
	Con cuál trabaja ahora? Haciendo qué?
	Trabajan juntos con organisaciones del gobierno? ☐ No ☐ Sí, con cuáles?
	Con cuál exportador trabajan?
	Están asociado a una cooperativa de mayor grado (segundo/tercero) y porque?

COOP-Members

La comunicacioon de la gerente con los socios, lo considera Excelente muy bueno bueno normal malo
What is the attitude of your members towards the cooperative? Do you feel they are strongly emotional bound?
Coops view of poverty of its members
➜ in my research all declared themselves as poor within the village/community. Why do you think you are poor? What have the rich what you don't have? What makes a person rich? What indicates that somebody considers himself as poor?

6

<u>Certificación</u>

	Cuál certificación tiene?
	Quien decidió cuál certificación?
	Quién les ayudó sacar la certificación?
	Quién financió? % coop-ONG...?

Porcentaje y volumen del café vendido en 2006-7

Typo	% del total	Volume qq
organico		
Organico-FT		
conventional		
Conv-FT		

Certification	Desde cuando certificado	Costos para sello/membrecía
EU		
NOP		
JAS		
OCIA		
BIOLATINA		
IMO/NATURLAND		

- Quién les certifica? Nombre de la persona responsable y costos de certificación?

OCIA		
BIOLATINA		
CERES		
BCS		

	Cómo paga para la certificación? Por la organización o el sello o ....?
	La cooperativa paga un precio fijo o un porcentaje relacionado al total de la venta de café?
	Cómo y cuánto pagan los productores? Por qq café? Cuánto pagaron en 2006-7?
	Para cumplir con los estandares de la certificación, los productores necesitaban de invertir en edificios, maquinaria o otra cosa? COSTOS???
	La cooperativa necesitaba construír algo nuevo o comprar maquinaria para cumplir con los estandares de la certificación? COSTOS???
	Tiene una estimación cuánto son los costos de la control interno de la certificación?
	Qué hace con los costos de la control interno? El productor lo paga, la coop?

7

	Cuál es el premio que les paga por café organico?
	Cuánto cuesta un taller para productores relacionado a la certificación? Cuántos son los costos totales para entrenar productores en la certificación en un año?

Relacion con los certificaciones

	Los compradores ofrecen a la cooperativa un adelante por el café? De que certificación? El adelante es cuanto % del valor del contrato? Suficiente?
	En los últimos 5 años, la venta directa aumento? Porque? Nuevos contactos, compran más..?
	A quién vendieron antes de obtener la certificación?
	Los miembros de la coop participaron en eventos relacionado al comercoii justo, orgánico fuera de Nicaragua? Quién financió? Grupos de productores también salieron?
	Cuál dice que son los beneficios más grande de la certificación orgánica? Por los Productores........y la coop
	Cuál dice que son los beneficios más grande de la certificación comercio justo? Por los Productores........y la coop
	Recibieron apoyo institucional de los certificadores y compradores certificado relacionado a consejos, talleres, información por dessarrollo de alta calidad, estrategias de venta, exportación, desarrollo de marca? ☐ No Sí, especifique
	Recibieron consejo/talleres tecnicos relacionado a la certificación?
	Hubo un apoyo de la certificación con crear connecciones con compradoras y cadenas de comercialización alternativas? Especialmente del comercio justo?

Qué transparente considera las		
	OF	**Comercio justo**
Reglas/estandares		
Contratos		
Flujo de información		

	Quién decidió como gastar el premio del comercio justo? Por voto? Cuándo deciden?
	Como utilizarón la premia del comercio justo en 2006-7?

8

Los otros años (desde 2002)?
Qué piensa del aumento del precio mínimo del comercio justo? Está suficiente de cubrir los gastos de producción? También suficiente para invertir?
Hay cosas que le gustaría que se mejora relacionado a las certificaciones?
Tienen un programa esecial para promover las mujeres?

SWOT FROM COOP VIEW

Cuál de las siguientes factores cree usted tiene la mayor influencia positiva en el desempeño del negocio de la cooperativa? Tecnología   Economía   Politica Legislacion   Cultura   Demografía
Cuál de las siguientes factores cree usted tiene la mayor influencia negativa en el desempeño del negocio de la cooperativa? Tecnología   Economía   Politica Legislacion   Cultura   Demografía
Cree usted que el ano pasado el desempeño de la cooperativa fue Excelente   muy bueno   bueno   normal   malo
Por favor, si piensa en el exito que tiene de la cooperativa, cúales de los siguientes factores son más importante? (Da un orden) _ Pago a los productores _ Desempeno en el Mercado /Venta/contactos con importadores _ Desempeno financiera _ Relaciones con los socios

Puntos fuertes:
1. Dónde piensa son sus puntos fuertes de la cooperativa?
2. Cuáles actividades de la cooperativa funcionan excelente?
3. Cuales son sus logros más grandes en los últimos 5 años?
4. En que está basado el éxito de su cooperativa?
5. Cuáles caracteristicas hace su cooperativa mejor que las otras?
6. Que dicen los otros (compradores, importadores,...) son sus puntos fuertes?

Puntos criticos
1. Dónde vea cosas que se deberían mejorar en la cooperativa (produccion, mercadeo)?
2. Cuál son limitaciones que encuentra en la cooperativa que hace más dificil lograr lo que quieren como cooperativa?
3. Hay cosas que critican los socios o compradores?
4. Y que piensa en hacer para evitar situaciones dificiles?
5. Que pasara con la cooperativa si los precios de café bajan otra vez?
6. Hay importadores que no quieren comprar de su cooperativa y porque?
7. Que son los recursos relevantes que les falta para tener más exito?
8. Podría uno de sus puntos críticos poner su cooperativa en peligro?

→ de todos que hablamos, cúal dice que son sus puntos más fuertes/debiles?

9

Amenazas del ambiente /Riesgos
1. Dónde vea riesgos de las cooperativas?
2. Habían grandes y significantes cambios en el sector cafetlaero?
3. Cuál son problemas que tienen las cooperativas?
4. Vea la competencia como un problema?
5. Hay nuevas reglas/estandares que pueden afectar el mercado del café?
6. Salen o entran socios en cooperativas? Y programas de certificación?
7. Cree que el cambio climatico afectara la producción? En que sentido?

Oportunidades
1. Dónde piensa son los oportunidades que tiene su cooperativa?
2. Cómo piensa se va a dessarollar el mercado de café?
3. Vea un chance de bajar los costos que tiene la cooperativa?
4. Hay posibilidades para nuevas relaciones ej con importadoras?
5. Podría mejorar la calidad del café vendido sin costos demasiados altos?
6. Hay un chance de pedir mejores precios por su café?
7. Ha pensado en amplificar sus servicios/ productos?
8. Había apoyo del gobierno en los últimos 5 años y especialmente desde 2007 relacionado a política, crédito, certificación, proyectos, talleres?
9. Hay eventos locales que benefician su cooperativa?
10. Participa en la taza de excelencia? Porque no?

10

Guiding Questions for Semi-Structured Farmer Interviews

Training/Capacity strengthening + Diversification
- Do farmers apply what learned in trainings? Reasons for answer
- Do farmers invest in their coffee plantations to improve conditions and quality?
- How are other cropping areas managed (conventional with chemicals, traditional without)

Certification
- What do you think about the certification? Why do you have it?
- Most important benefit from FT / OF?
- Did economic security increased through certification?
- Have been already controlled by an external certifier? How was experience with that?

Coffee marketing
- How do you sell your coffee to the cooperative? Knowledge of the value chain?
- What are experiences with intermediaries? Why do they sell to them?
- What would they consider a fair price?

Cooperatives
- What are advantages and disadvantages to be organized?
- How is the payment for coffee? Timely? Transparent?

Financing
- Which organizations provide credit?
- What is preferred and why?
- What do you do if you urgently need a credit?

Poverty
- Development of their living conditions (improvement or worsening), knowledge (expansion or reduction) and rights (participation or deprivation)
- Strategies which could reduce it?

General labour
- Division of labour - men/women?
- Compare work burden ORG, OFT, CONV of men and women
- Scarcity of seasonal labour esp. in harvesting a problem?

Gender/Women
- Decision-maker and decision process in household
- Decisions about expenditures / investments?
- Perception of gender roles in coffee farming?
- Representation of women in coop / why are women not registered in coop?
- Perception how life changed after ORG and OFT certification?

— Appendix D —

Topics for Focus Group Sessions

<u>**IMPACT INTERVIEWS:**</u>

Find out certification impacts on
(unstructured interviews with sometime guiding questions):
- experiences with certification and internal control system
- income situation – changes?
- living conditions
- coffee marketing
- investments on farm
- capacity strengthening (technical, knowledge, organisation)
- access to financing
- poverty
- gender ➔ labour division, decision making
- experiences with being coop members

<u>**SEASONAL CALENDAR**</u>

a) Which labour activities in coffee in each month?
b) Labour activities for other crops
c) Amount of off-farm labour per month
d) Income distribution in each month
e) Months of food scarcity and abundance
f) Months with low/high expenditures
g) Credit / debt
h) Low and high prices for food
i) Health

<u>**GENDER**</u>

General labour & decision-making
- Division of labour among men and women
- Work burden for CONV, ORG, and OFT cooperative members - men and women (elaborate daily timetables)
- Decision-maker and decision process in household
- Decisions about expenditures / investments

Women
- Perception of gender roles in coffee farming
- Representation of women in coop.
- Why are women not registered in coop.?
- Perception how life changed after ORG and OFT certification